AF677289

THE BAMBOO ROD

THE MACMILLAN COMPANY
NEW YORK • BOSTON • CHICAGO
DALLAS • ATLANTA • SAN FRANCISCO

MACMILLAN AND CO., LIMITED
LONDON • BOMBAY • CALCUTTA
MADRAS • MELBOURNE

THE MACMILLAN COMPANY
OF CANADA, LIMITED
TORONTO

THE

BAMBOO ROD

AND HOW TO BUILD IT

by Claude M. Kreider

New York

THE MACMILLAN COMPANY

1951

COPYRIGHT, 1951, BY THE MACMILLAN COMPANY

All rights reserved—no part of this book may be reproduced in any form without permission in writing from the publisher, except by a reviewer who wishes to quote brief passages in connection with a review written for inclusion in magazine or newspaper.

First Printing

PRINTED IN THE UNITED STATES OF AMERICA

25499

799
K

¶This book is dedicated to the memory of Robert Crompton, master rod builder and lover of the craft. A marvelous workman, with the deep, inquiring mind of a scientist and engineer, he was never satisfied with less than absolute perfection in his work.

¶For many years he gave unstinting help and advice to both amateur and professional rod builders throughout the country. And for his counsel and advice the author shall always feel deeply grateful.

INTRODUCTION

There is something about the marvelous action of a good split bamboo rod which appeals to every experienced angler and grips his imagination. The rod possesses *life,* a springlike power and strength coupled with a smooth resistance when subjected to strain. And no other material used for rod making seems capable of producing all these desirable characteristics.

It is my earnest endeavor in this little work to better acquaint brother anglers with something of the nature of bamboo and its use in rod making down through the years; to explain some of the details of the fine craftsmanship employed by our best rod makers to produce the splendid bamboo rods we use today, and thus, in touching upon their practices and technique, to present an over-all picture of the art employed by the best professional rod makers.

This information will in turn, I sincerely hope, help the interested amateur craftsman who has always wanted to build his own rod to reach the point where he will actually plan the work. The necessary equipment required by the home rod maker is simple and inexpensive. The labor is light, and utterly absorbing. And a high degree of mechanical training is wholly unnecessary—as I can testify in my own case.

Once you have started the work and find that beautiful rod of your dreams actually taking shape in your hands, you will become a slave—and a willing one—to your work bench. You will work long hours into the night, and mumble strange figures to do with rod calibrations through the day. Family and friends may wonder and look at you askance. But never mind, *you* are building a fishing rod.

Then comes that unforgettable day when you breathlessly assemble the new rod on the bank of your favorite stream. At last it is a reality, a thing of life and beauty, delicate, pliant, yet powerful and ready to serve you faithfully for years to come. This is *your own rod,* and this will prove the sweetest moment of your angling career.

Grateful acknowledgment is made for data provided by Mr. Nathaniel Uslan of the Uslan Rod Company, Mr. Lew Stoner of the Winston Rod Company and others. And particularly I wish to acknowledge the advice and assistance of Robert Crompton, a master rod builder who, through the years, helped guide me through the intricacies of home rod building.

Finally, I am indebted to Mr. Larry Tong and my son, Peter Kreider, for the drawings used for illustrations.

CONTENTS

CHAPTER 1

THE USE OF BAMBOO IN ROD BUILDING

Even the experienced angler may be surprised to learn that his favorite bamboo rod, that beautiful and strong yet pliant implement, comes from a genus of grasses. But so scientists classify this marvelous and useful material, giving it the name *Bambusa.*

There are some hundreds of varieties of bamboo known throughout the world, principally in the tropical and semi-tropical regions of the Eastern Hemisphere. It also occurs in some sections of the United States, but this domestic variety seems of inferior quality and little, if any, has been found suitable for rod building.

Asiatic bamboo of the best quality, from carefully selected canes, is of truly marvelous structure. The fibres are strong, tough, resilient and resistant to fatigue. And these little "backbones of power" extend the full length of the cane, except where they cross and interlace at the leaf nodes. In the well-constructed rod these few weaker spots, which might be likened to small knots in other woods, are wisely spaced in each rod section, and each will be opposed by solid fibre in the other splines which make up the stick.

Most good Asiatic rod bamboo is referred to as "Tonkin," probably because it comes from Tonkin Province, in southern China. Another variety of bamboo, possibly from the same general district, is designated by some rod manufacturers as "Tsinglee." In any event, the best rod canes are usually from one and one half to two inches in diameter, being cut from the butt sections of straight, carefully selected stocks which are about eight years old. These usually come to us in lengths of eight or nine feet.

Well-seasoned rod canes are of a pleasing light yellow or tan color, very smooth and straight, with wide-spaced leaf nodes barely protruding above the surface. Underneath the hard, exterior enamel, which seems to be a sort of natural, protective varnish, and which has no strength in itself, lie the hundreds of tiny, hard, siliceous fibres, which impart to the fishing rod its extraordinary power, strength and smooth "live" action.

The finest of these fibres occur just under the enamel. They become increasingly coarser toward the inner wall of the cane, and separated by layers of light colored pith. Thus, only a thick-walled cane which has a considerable area of the dense outer fibre is suitable for rod work if there is to be the densest of fibre structure extending in to the core of the rod.

The development of today's fine bamboo rods has been one of slow evolution dating back to early in the nineteenth century, when a few British craftsmen first experimented with "rent cane, split and glued," to replace the hickory, lancewood, bethabara and other hardwoods then in use. And we can well understand the delight of these pioneers in discovering a material so superior to hardwood, after centuries of shaping up hardwood billets into rods of various types.

The old hardwood rods were heavy for a given type, prone to warp and take a "set" even after moderate strain, and of slow, logy action. The trout and salmon rods of those days were indeed fearsome implements, an ordinary "light" trout rod being eleven or twelve feet in length and weighing probably from twelve ounces to a pound. A nice, two-handed salmon rod might measure fifteen feet or more and weigh around *two* pounds. Certainly the British sportsmen of those days must have been men of mighty brawn.

In the first experiments with bamboo it was used only for rod tips. Three- and four-strip construction was employed for many years before the conventional six-strip or hexagonal design we know today came into general use. It is quite probable that for many years those early pioneer rod builders worked only with Calcutta bamboo, long before discovery of the Tonkin variety which has proved far superior.

Calcutta canes come from India and may still be found in almost any tackle store today for use in their natural form as simple "fishing poles." They are tough, flexible and remarkably strong and many thousands of these, with the slender tips cut back and mounted with the proper fittings, are used for surf and general salt water fishing.

While the fibres of this bamboo are tough and straight, the canes are of small diameter, having thin walls and many rough leaf nodes. Thus they will not provide sufficient good material for splines to make large rod sections. This is the reason, perhaps, why the early craftsman attempted only light tips of split bamboo. But later, possibly with a greater supply of canes from which the best could be chosen, the British craftsmen evolved trout and salmon rods that were "all bamboo."

These were probably still of three- or four-strip construction. There is little literature which follows the consecutive steps of the development of the art through those years, but there is no question that complete rods of six-strip design were in use by British anglers by the year 1850. It is not difficult to imagine the keen sportsman of a hundred years ago trying out his first new "split-cane" rod on the great salmon and sea trout of a Scotland loch, then hurrying back to his rod maker in London or Glasgow to place an order for another of these new and delightful implements.

American craftsmen, of course, were not slow to recognize the merits of the all-bamboo rod, and we have good authority for the statement that the first of these were built in the United States by Charles F. Murphy of Newark, New Jersey, and placed on sale in the early 1860's. These rods were of four-strip construction.

The first six-strip American-made rods were placed on the

market by H. L. Leonard of Bangor, Maine, in 1870, thus marking the genesis of the beautiful bamboo rods we use today. Rod design has changed greatly since that time, the demand increasing for fly rods of lighter weight, shorter length and greater power. Bait casting rods are longer with more whip and finer action than those of the past, but there is no finer workmanship to be found in the best rods today than that in the old original "Leonard."

There was probably little change in basic rod construction for many years, the bamboo being split by hand, the strips carefully straightened then planed down, a shaving at a time, until each became a perfect triangular section, tapered accurately to a thousandth of an inch tolerance to fit its neighbors and so form a rod section. This is the method used by the amateur rod builder today, as will be described in later chapters. The commercial production of rods, of course, has gone far beyond simple handwork methods, with special precision machinery, improved technique, better bamboo and stronger, waterproof glue insuring far better rods than were ever possible in the days of Leonard and contemporary artists.

Improved design—various combinations of taper—is giving us more in the various types of fly, bait casting and spinning rods than ever before. The tapers of a modern fly rod are worked out to exact formulas, and literally hundreds of types, from the little two-ounce wand to a powerful salmon rod, are available on the market. The amateur builder, then, can profit from this scientific development in rod design and accurately copy any rod he wishes. How nearly he may come to matching the masterpiece of our best commercial builders will depend upon the excellence of his workmanship.

The development of tournament casting in recent years has led to many new rod designs. Scores of the leading performers in this highly scientific and interesting game have contributed new and valuable—sometimes revolutionary—ideas as to the action required to secure the ultimate in a rod of given type.

Progressive manufacturers have not been slow to recognize many of these suggestions and have incorporated them in the best of their products, thus offering a wider selection to the angler than ever before.

But quantity production as it must be practiced by the large firms today can, after all, supply only the standard types of rods most in demand. Those in the higher price ranges will be beautiful implements, superbly fashioned and finished, and well worth their cost. The less expensive rods, even though almost entirely machine produced, with little or none of the expert hand finishing used on the finer rods, will also give faithful service.

But the tournament expert and the angler who, admittedly, is a "rod crank" will often pass up the finest rod in the tackle store and go into serious consultation with one of the few expert rod craftsmen, still to be found, whose limited production will permit any variation in rod action desired. These masters of the art delight in satisfying the discerning buyer who appreciates fine work, and will go to almost any length to please him. It is seldom necessary actually to build to order a special rod. The buyer tries various rods of the general type desired, explains any changes that seem necessary, and the expert craftsman can usually meet these requirements by carefully calibrating and selecting a set of rod sticks from his seasoned stock.

Fishing rods for a specific use are being built lighter than ever before, with angling now a highly scientific sport, and marvelous results are obtained with tackle so light it would have been scorned as silly fifty years ago. One can have a tiny, seven-foot trout rod now, weighing a mere two ounces, which will cast a fly fifty feet with ease, and in expert hands can battle to a finish a trout of ten pounds. The bait caster, whose game is the rough and tough black bass, can buy a slender six-foot wand which will whip a plug out a hundred feet and tame the toughest old largemouth with its smooth, resilient power.

In the field of salt water fishing, scores of fishing clubs have adopted tackle rules which require rods built to exact weight specifications, thus placing a premium upon the angler's ability and employing to the utmost each ounce of bamboo in the stick. Perhaps an all time high, in this respect, was registered by William Pigg at La Paz, Lower California, when in 1948 he took a great marlin of 175 pounds on "3-4" tackle. The line tested only nine pounds, and the five-foot rod tip of bamboo weighed just *three* ounces.

Turning to fly rods for the ultimate performance in that field, we find that a distance-tournament rod, of nine and one half feet, weighing about six ounces, has cast a fly more than 180 feet in expert hands. Thus, these few instances selected from many examples of superb rod performance indicate the possibilities of this marvelous wood, expertly fashioned into a fishing rod.

Whether we can expect further improvement remains purely a guess. Certainly the angler will be ever-hopeful. If, after reading the following pages, he decides to undertake home rod building he will find it an absorbing pastime, with each new rod from his work bench more nearly approaching his idea of perfection.

CHAPTER 2

METHODS OF ROD CONSTRUCTION

Many anglers believe that a really fine bamboo rod can only be constructed of splines shaped from strips actually *split* from the cane. This premise, while basically sound, has probably been handed down from the days when rods were entirely handmade and, of course, the strips split out by hand.

Splitting does follow the grain and so maintains the long bamboo fibres the full length without break, and this method is used today by perhaps a majority of manufacturers. The usual method is to force an entire cane into a set of stationary knives so adjusted as to separate it into the desired number of strips. These are then sent through a shaping machine, somewhat similar to those used in planing mills for cutting moulding. High-speed, revolving cutters thus bevel the splines, and various adjustments determine their taper. Several cuts may be taken from each strip, the last a very fine one to assure a smooth and accurate surface for exact joining before the gluing operation.

Other firms produce fine rods, probably equally good, by sawing out and shaping the splines in a pair of tiny circular saws set at an angle to bevel them accurately for exact fitting. While this method may be frowned upon by proponents of the splitting technique, it certainly cannot be particularly harmful, since the bamboo fibres do extend, fairly straight and true, the length of the cane. Regardless of the method used for shaping the rod splines, the essential thing is accurate beveling to secure exact fitting at all points for gluing.

Gluing of the strips, or "bonding," as the professional terms it, is the most important operation in rod building, and perfect results can be obtained only if all the strips of a rod section join perfectly at every point. Under severe strain, as in ordinary use, there is a strong tendency for the strips, or splines, to slide past each other. When the rod is arched the upper surface is actually stretched and the lower surface is compressed. This action exerts a terrific strain upon each glued surface, particularly the central one which evenly divides a rod of six-strip construction.

There will be perhaps a thickness of only .001 or .002 inch of glue between these surfaces, much less than the thickness of a human hair, so we can appreciate the importance of perfect bonding. Better glues are used in present-day rods than ever before. The "animal" and "fish" glues generally used by earlier rod craftsmen required an exact and complicated technique to secure even passable results. The workroom and glue temperature had to be carefully controlled, the work done speedily and the glued sticks given a long period of careful seasoning afterward. Also, these glues were highly susceptible to moisture, heat and even to changing humidity.

The older rods, even of the highest grade, were prone to take an unaccountable "set" on a hot, humid day, to feel slow and lifeless, so easily was the glue affected. And, even when well varnished, such a rod might actually separate at the strips if left in a damp bag, without a chance to dry, for a day or two.

Today the casein adhesives and the even better synthetic resin glues of many types are in general use, and the fine rods we now enjoy will be so strongly bonded that the wood itself will probably split before the glue joint separates. Some of the resin glues are suitable for use only under high pressure and great heat, while others, easily applied with a simple technique, are ideal for the home rod maker's use.

The casein glues are highly water resistant and the resins practically waterproof, a tremendous advantage when we consider that all rods in use are subject to moisture and often are unintentionally left damp throughout a night or longer. I have trimmed a short section from a rod stick glued with one of the resins later described, have actually *boiled* it in water, then had difficulty in separating the splines with a chisel.

As mentioned in Chapter 1, the wall thickness of the cane will limit the size of the rod stick. In even the best canes the area of close-packed outer fibres will seldom exceed a thickness of a quarter inch; and often it is less. This diameter, then, extending from the exterior of the rod to its center will give us half the stick's diameter. Doubling this we find that a half inch of over-all diameter makes possible only fly and bait-casting rods, and the lighter types of salt water rods in *single built,* standard six-strip construction.

Since larger rods also must incorporate the finest fibres in to the core, if they are to be tough and serviceable, various methods of lamination have been evolved in which as many thin strips of fine fibre as necessary are literally *welded* by the modern glues into one rod stick.

Thus was evolved *double built* construction, still employing the conventional six-strip design. Two flat strips of bamboo, with all the interior pith and coarser fibres shorn away, are then glued face to face. Then these six laminated sticks are beveled in the usual way and glued up. There are twelve glue faces now, to be faultlessly matched and properly bonded, requiring extreme care and accurate workmanship.

In larger sticks, such as those made up into heavy tackle rods, used for swordfish and giant tuna, even *triple building* is resorted to if the rod is to stand the hard service demanded of it. Careful fitting and gluing of these eighteen bamboo splines calls for the very highest type of workmanship in the rod building craft.

Another type of lamination is used without beveling the splines, with square strips, or with those hexagonal in cross section being glued up to form a rod that is square. Other makers using this construction finish the rod trapezoidal in cross section; and still other laminated rods have two sloping sides, with a wide, flat base opposed by a curved, oval surface.

It would be interesting to have the unbiased opinion of competent engineers as to the merits of these various forms of construction. But even then we would probably learn little from the practical standpoint. How the rod performs for the angler is what counts, and we find experienced users swearing by each of the rod types mentioned. So it would seem that each of the several designers has arrived at about the same point, though by a different method.

While the conventional six-strip rod has been standard in the lighter calibres for many years, several makers have departed from this design, always with the hope, it seems, of finding something still better. Thus, there is the square, or "quadrate" rod, made of four 45-degree splines, and it is commended highly by competent anglers who have used it. But the square rod has four equal surfaces, with two "flats" opposing during the stress of use. The six-strip, also, has opposing flat surfaces, though they are braced by the two central apexes, one on each side. Now, it would seem quite logical to the layman that if one apex of the rod opposed the flat side upon which the guides are mounted, since this is the direction of arc—and strain—greater stiffening effect, thus strength and power, would be secured from a given amount of bamboo.

Perhaps the possibilities of this design may have occurred to thoughtful rod makers throughout the years, some of whom did try eight-, and possibly ten-, and twelve-strip construction. But always, of course, an even number of strips gave them opposing flat surfaces. The answer was logically and simply the use of an odd number of strips in which an apex must always oppose a flat side of the rod.

Three-strip construction was impractical, the use of seven or nine narrow splines involved more intricate and careful shaping, so five-strip design seemed logical. This was experimented with by a master rod builder of the old school, Robert Crompton, many years ago, and he painstakingly worked out many sets of tapers for its use, also designing steel planing forms for hand shaping the splines and a revolutionary milling machine for speed production.

The five-strip rod definitely has unusual merit, as I can testify after building very many of them to Robert Crompton's charts, and others of experimental tapers. Beautiful five-strip rods of various types are now being produced commercially by the Uslan Rod Company of Spring Valley, New York. Perhaps other rod builders supply them to order for a limited clientele; and certainly a great many amateur craftsmen are enjoying the novelty of five-strip construction, with the satisfaction of producing very fine, useful and unusual rods of the marvelous bamboo fibre.

Another definite advantage of five-strip construction is that the glue line is off center, the working strain being upon solid wood (note Fig. 2 in the illustration) rather than upon a continuous glue line in the six-strip, Fig. 1. Thus, the slipping, shearing tendency which must be present in the six-strip is entirely absent in the five-strip pentagon. My own layman's comparison is not intended to detract in the least from the merits of the six-strip rod, whose long and honorable record speaks for itself. It is offered only to stimulate interest in the amateur rod builder, who will derive much pleasure from constructing and comparing rods of both types.

Before the final shaping, rod canes must be slowly and carefully seasoned to remove both the natural sap in the wood together with moisture which may have been absorbed from the air. This change probably takes place to some extent during the months of shipping and storage by our importers. But rod manufacturers give it a final scientific and carefully controlled treatment to further remove any moisture present.

The full canes, and sometimes bundles of roughed out strips, are placed in special rooms, or dryers, with temperature and humidity under exact control. The amateur workman, lacking such scientific means of curing can, nevertheless, order the best grade of canes, well cured, and store them in a warm, dry place until needed.

The curing process above referred to should not be confused with the so-called "heat tempering" of bamboo, used by some makers to harden the bamboo structure still more and add a certain stiffness to the finished rod. Perhaps each maker who so tempers his rods uses a different and sometimes secret treatment, but the basic method is to subject the wood to a high degree of controlled heat.

This possibly does harden the fibres, and it certainly turns the wood to varying darker shades, from light tan to dark brown, depending upon the temperature. But there is a critical point when extreme heat will definitely injure the bamboo structure and render it brittle. Such a rod may break off short under stress.

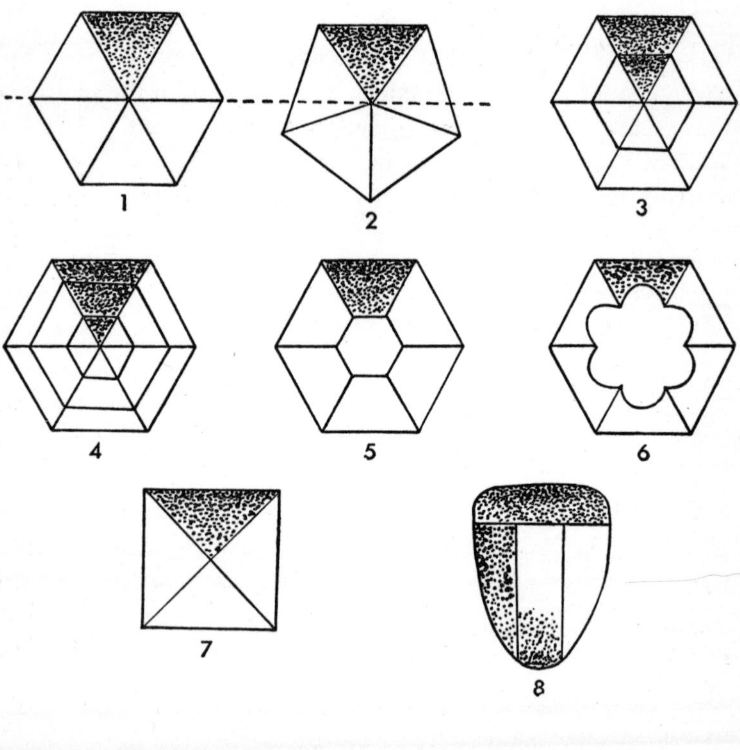

Other manufacturers frown upon heat tempering and still produce fine rods from bamboo in its natural state. So the matter of heat treatment is one subject to argument, with proponents on both sides. The home craftsman will find it interesting to experiment with a mild heat treatment, as described later, and he then can form his own opinion.

Just as our best rod makers disagree upon the efficacy of heat tempering, so their ideas diverge in regard to "solid built" against hollow rod construction. It would seem that a rod of solid fibre to the core would be stronger and of better action than one in which the inner apexes of the splines had been removed to leave a hole through the center. But this inner portion which is removed is the coarser structure, and its loss can be compensated for by a slightly larger over-all diameter of the rod, containing more of the best outer fibres of the bamboo.

This idea is carried out in the unique "fluted hollow" rod built by the Winston Rod Company of San Francisco and has been

◀ VARIOUS FORMS OF ROD, CONSTRUCTION, DOTTED AREAS SHOWING HOW DENSEST FIBRES ARE NEAR OUTSIDE OF CANE

1 *Six-strip. Dotted line indicates center of strain is on glue line.*
2 *Five-strip. In this pentagon center of strain is on solid wood.*
3 *Double built construction, employing 12 thin strips of bamboo in order to utilize only the dense fibre area of the cane.*
4 *Triple-built, going a step farther than in the double-built, and using 18 strips of bamboo.*
5 *Hollow-built. Removal of inner apexes of the strips eliminates only the coarser grain, and reduces weight of rod.*
6 *Fluted-hollow rod. Still more of the inner wood removed, and only the finest of the outer grain retained. The longer apexes at glue lines still insure strength in the glued stick.*
7 *Square construction, using triangular-shaped strips.*
8 *The flat-strip method of lamination, still utilizing only the dense fibre area of the bamboo.*

developed by their expert designer and shop manager, Lew Stoner. He points out that in this construction the bending stresses do not occur entirely on the convexly and concavely bent sides of the rod, but are relieved by the lateral expansion and so distributed equally upon the entire rod circumference. And, whether or not we mere laymen can fully understand this interesting theory, we can appreciate the basic soundness of the fluting principle in building a hollow rod. This removes more wood from the apex of each spline (note Fig. 6) than would a flat cut, yet leaves a wide glue surface at each side. Fig. 5 shows the usual tubular construction, easy to accomplish by the home rod builder who wishes to experiment with hollow building.

This practice, of course, is used only in building the larger rod sections. It can well be employed in fly rod butt sections and in middle joints to a moderate degree. Small tips cannot well be hollowed, or there would be no wood left to work with.

The saving of weight in a tubular rod is actually of little consequence. However, it does permit the angler who uses overall weight of a stick as a means of defining its power, when he displays his ultra-light, hollow-built favorite, to expound upon the fine action obtained with so few ounces of bamboo.

There are certain mechanical advantages in moderate tubular construction for the amateur rod builder, which will be described in detail in Chapters 8 and 9.

CHAPTER 3

ROD TAPERS AND TYPES

As the amateur workman progresses in the art of rod building he will develop a keen interest in studying rod action, in applying his micrometer to various rods, both good and "bad," in an effort to determine just how the various tapers affect rod performance. And only micrometer readings in thousandths of an inch will tell the story. Crude measurements, such as those possible with an ordinary caliper, will not do at all.

So critical are the calibrations of a fly rod, it will at first glance appear to have a constant taper from hand grasp to tip. But this design will actually produce only a rod of "soft," unresponsive action, without that indefinable quality we call "backbone," so essential in a fly rod. A desirable rod, though not necessarily stiff, will seem to possess life, respond quickly to the angler's efforts, then return to its normal position, without a wobble and a slow series of vibrations.

Fine material and workmanship will, of course, contribute to the rod's performance, and these are basic necessities. But the secret of its action will be in the varying tapers employed throughout its length. In short, the amount and quality of bamboo and the workmanship will indicate what the rod is *capable of*. The taper will determine *how* it will perform.

In general, the method used in producing the fine fly rods of today is to employ a slow taper, perhaps a reduction of .010 inch for each six inches of length in the lower half or two thirds of a rod section, then to increase the taper more rapidly toward the end. Just where this increased reduction starts and its severity will affect the rod's action to a greater or less degree.

Unfortunately, there are no accepted terms defining rod action which will convey to all anglers the actual performance of the many types of fly rods. Thus, the conventional dry fly rod usually indicates a stick having a fairly stiff lower section with a comparatively light, flexible tip. This is often called a "fast" rod. It throws a narrow loop of line and permits greater accuracy with the dry fly than a "slow" rod, which flexes more evenly through its length. A rod of this latter type is often considered a good rod for wet fly fishing.

Somewhere between the two actions described there may be a modification which does insure a good wet fly rod, but these days anglers demand a fast rod for all types of fly fishing, and the good tip-action rod seems to fill the requirement, so long as its tip will not undergo too severe strain in playing a fish and making long casts. Thus, if the amateur rod builder wishes to

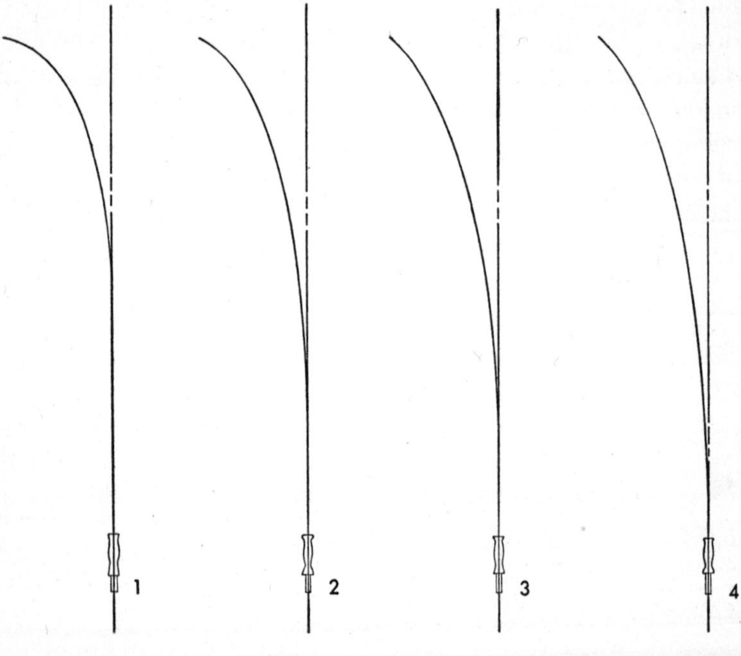

copy a good dry fly rod and enjoy its fine, fast action, but add a bit more strength for distance casting, he need only add a few thousandths of an inch to the tip calibrations.

In recent years we have heard much of the "parabolic" rod which, roughly described, arches well down toward the grip when sufficient stress is applied. The taper will be more constant than in the dry fly rod, with a heavier tip and possibly a lighter butt section than in a dry fly rod of similar length. As more load or working strain is imposed on this rod the heavier tip begins to "shrug" off the work and distribute it farther down. Since the butt portion is light in comparison to the tip, it more quickly responds with an increasing arc, thus soon throwing the entire rod length into play.

The parabolic type thus utilizes all the bamboo to best advantage as more work is demanded of it. But only the expert caster, adept at proper timing of the back cast as more line is handled, can develop the full potential power of the parabolic rod. Its general action is exemplified in the tournament caster's distance fly rod with which such great distances are covered. The general design of the parabolic principle can best be noted in the typical spinning rod which, even with moderate effort and a very light lure, will arch noticeably down to the cork grip.

A short study of the first two rods charted on page 21 will indicate the slight changes in calibrations at various points which

◀ ILLUSTRATING FLY ROD ACTION, DESCRIBED BY THE COMMON TERMS IN GENERAL USE

1 *Dry fly, or tip action.*
2 *Modified dry fly action.*
3 *Semi-parabolic action.*
4 *Parabolic action as developed in the spinning rod.*

so vitally change rod action. No. 1 is a very fine dry fly rod and might be considered of "modified" type, though not with extreme tip action. No. 2 develops very much more action down through the butt section, yet is only slightly smaller up to 36 inches above the lower end. It increases in size over No. 1 from 42 to 54 inches; thus it more quickly puts the lower portion to work. And the tip, considerably heavier, adds to this tendency to distribute the work at once down the rod.

The study of fly rod tapers and resulting rod action is an exhaustive one with which the finest rod designers are constantly engaged. So until the amateur has progressed far into the work he had best be content to copy the proved designs of rods which he likes.

Tapers are less critical in rods of the bait casting type, a fairly constant reduction in size providing a pleasing action and one capable of handling the heavier lures, so long as "enough wood" is used to withstand the constant strain of casting these.

The heavier types of salt water rods will probably not interest the amateur builder at first, for these involve the difficult technique of laminating many strips, as earlier discussed. Lighter sticks, however, such as the "3-6" and "light tackle" rods, used on the Pacific coast, and weighing four and six ounces respectively, without the butt, and those of somewhat similar size, favored along the Atlantic coast, are easily possible. These rods will require the best of thick-walled cane to insure the necessary strength without double building them.

While a discussion of over-all weight is necessary in describing salt water rods, since they must conform to a weight and length standard set up by the various fishing clubs, it is misleading to attempt to define the possibilities of fly and bait casting rods in this manner. Rod weight means nothing in relation to rod action, since the grip, reel seat and other fittings will vary so widely in type, and so affect the rod's over-all weight but not its action.

A hollow-built nine foot fly rod weighing perhaps only 4½ ounces may prove as powerful as a stick of similiar length weighing a full ounce more in solid construction. Where over-all rod weight is listed in the following rod charts, it is merely for comparison, since we are accustomed to this description of fishing rods, and no better method can readily be offered.

Another much discussed point in fly rods—of a matter which is actually of little consequence—is rod balance. The angler wants his ideal fly rod actually to balance, perhaps, at a point three inches forward of the hand grasp. Such a rod may "feel fine" when whipped through the air without being fitted with reel and line. But a reel weighing from five to seven ounces will radically change the point of balance; when line is passed through the rod guides that mythical point will again be changed. Since the rod is built, primarily, to develop a certain definite action there certainly can be no shifting of calibrations which would actually change the balance point of the completed rod.

If a certain balance is desired in the fly rod the only practicable method of achieving it is to mount a heavier—or lighter—reel seat, and later to test the rod with reels of various weights. In general, it will be found that excessive weight at the rod butt—the reel—will slow rod action, exerting a kind of pendulum effect, and that a lighter one will seem to liven up the rod's response.

The above various points have been briefly touched upon in an effort to guide the amateur builder in deciding just what type of rod he will first attempt, with some idea, perhaps, of how it will perform. The following rod charts may serve as an additional guide, and for comparisons. The rods chartered are by various makers, chosen to provide a fairly comprehensive field of study, and include products of only a few of the many fine manufacturers of the "split bamboo," along with examples of the amateur's work, including the author's.

All calibrations are in thousandths of an inch, taken across the rod "flats" in the six-strip, and from flat to apex in those of five-strip construction. Slight variations or errors may be present in the commercial rods, due to the varnish coat over which calibrations were made, but these will be minor. The diameter of the lower twelve inches is omitted, since this could not be obtained with cork grip and reel seat mounted on the finished rod. In general the diameter of these rods at the butt and 6-inch stations may be assumed to be an increasing taper from that at twelve inches.

CHAPTER 4

ROD CHARTS

FLY RODS—SIX-STRIP CONSTRUCTION

9 FOOT, 2 PIECE

No. 1 By E. C. Powell Rod Company.

	End	*6"*	*12"*	*18"*	*24"*	*30"*	*36"*	*42"*	*48"*	*54"*	*Ferrule*
Butt			.400	.365	.345	.320	.300	.275	.255	.240	.234
Tip	.235	.225	.210	.195	.170	.155	.138	.125	.110	.075	

No. 2 By Edwards Rod Company.

	End	*6"*	*12"*	*18"*	*24"*	*30"*	*36"*	*42"*	*48"*	*54"*	*Ferrule*
Butt			.365	.355	.330	.315	.295	.280	.265	.245	.237
Tip	.240	.235	.215	.195	.190	.163	.150	.130	.100	.075	

This rod weighs 5⅝ ounces with solid butt and screw type reel seat.

No. 3 Has nice "modified dry fly" action.

	End	*6"*	*12"*	*18"*	*24"*	*30"*	*36"*	*42"*	*48"*	*54"*	*Ferrule*
Butt	.390	.390	.360	.345	.335	.320	.300	.275	.265	.250	.250
Tip	.250	.235	.215	.195	.175	.150	.135	.120	.100	.085	

No. 4 A very powerful rod of bass bug type.

	End	*6"*	*12"*	*18"*	*24"*	*30"*	*36"*	*42"*	*48"*	*54"*	*Ferrule*
Butt	.430	.430	.405	.385	.365	.345	.320	.300	.275	.265	.265
Tip	.265	.260	.245	.230	.205	.175	.155	.135	.115	.100	

Built with solid butt and fitted with sturdy metal, screw type reel seat, this rod weighs 6½ ounces.

FLY RODS—SIX-STRIP CONSTRUCTION

9½ FOOT, 2 PIECE

No. 5 By E. C. Powell Rod Company.

	End	*6″*	*12″*	*18″*	*24″*	*30″*	*36″*	*42″*	*48″*	*54″*	*57″*	*Ferrule*
Butt			.390	.375	.360	.340	.325	.300	.285	.275	.265	.265
Tip	.265	.250	.235	.210	.195	.175	.150	.135	.115	.100	.090	

A very powerful rod, but in hollow-built construction weighing only a bit more than 5 ounces.

No. 6 By Winston Rod Company.

	End	*6″*	*12″*	*18″*	*24″*	*30″*	*36″*	*42″*	*48″*	*54″*	*57″*	*Ferrule*
Butt			.425	.385	.360	.355	.345	.330	.315	.310	.305	.296
Tip	.300	.290	.280	.255	.230	.215	.205	.190	.155	.125	.110	

An excellent salmon and steelhead rod, more powerful than No. 5 and weighing about 6 ounces with fluted hollow butt.

No. 7 A dry fly rod built by the author, with which James Corbell won the 1947 National Accuracy Dry Fly Championship.

	End	*6″*	*12″*	*18″*	*24″*	*30″*	*36″*	*42″*	*48″*	*54″*	*57″*	*Ferrule*
Butt	.390	.390	.370	.360	.345	.325	.305	.290	.280	.270	.265	.265
Tip	.260	.245	.220	.195	.180	.160	.135	.125	.110	.090	.080	

FLY RODS—SIX-STRIP CONSTRUCTION

8½ FOOT, 2 PIECE

No. 8 By Winston Rod Company.

	End	*6″*	*12″*	*18″*	*24″*	*30″*	*36″*	*42″*	*48″*	*51″*	*Ferrule*
Butt			.360	.340	.325	.310	.295	.285	.270	.260	.250
Tip	.245	.230	.220	.195	.170	.155	.135	.110	.095	.080	

A fine, powerful, little rod, weighing 4½ ounces in fluted hollow construction, with standard locking type reel seat.

No. 9

	End	*6″*	*12″*	*18″*	*24″*	*30″*	*36″*	*42″*	*48″*	*51″*	*Ferrule*
Butt	.380	.375	.350	.325	.300	.285	.270	.250	.230	.220	.218
Tip	.215	.205	.185	.175	.160	.140	.120	.105	.085	.078	

This is a nice, light-action rod, but in solid-built construction weighs as much as No. 8.

FLY RODS—SIX-STRIP CONSTRUCTION

8 FOOT, 2 PIECE

No. 10 By Winston Rod Company.

	End	*6″*	*12″*	*18″*	*24″*	*30″*	*36″*	*42″*	*48″*	*Ferrule*
Butt			.295	.280	.270	.250	.245	.235	.220	.215
Tip	.220	.215	.200	.175	.155	.140	.115	.093	.070	

No. 11

	End	*6″*	*12″*	*18″*	*24″*	*30″*	*36″*	*42″*	*48″*	*Ferrule*
Butt	.350	.350	.320	.290	.285	.270	.255	.235	.220	.218
Tip	.220	.210	.200	.185	.165	.140	.115	.100	.085	

FLY RODS—SIX-STRIP CONSTRUCTION

9½ FOOT, 3 PIECE

No. 12 Leonard Tournament Rod

	End	*6″*	*12″*	*18″*	*24″*	*30″*	*36″*	*38″*	*Ferrule*
Butt			.380	.370	.350	.330	.310	.300	.296
Middle Section	.300	.287	.278	.263	.240	.230	.220	.200	.187
Tip	.195	.177	.160	.145	.125	.095	.075	.070	

With heavy butt and light, flexible tip, this rod possesses both power and fine dry fly action.

No. 13 Powerful steelhead or brass bug type.

	End	*6″*	*12″*	*18″*	*24″*	*30″*	*36″*	*38″*	*Ferrule*
Butt	.420	.420	.400	.385	.370	.350	.335	.325	.328
Middle Section	.325	.300	.290	.280	.265	.240	.225	.220	.220
Tip	.215	.205	.190	.170	.145	.120	.105	.100	

FLY RODS—SIX-STRIP CONSTRUCTION

9 FOOT, 3 PIECE

No. 14 Powerful action. Heddon No. 17.

	End	*6″*	*12″*	*18″*	*24″*	*30″*	*36″*	*Ferrule*
Butt			.375	.367	.357	.350	.345	.343
Middle Section	.340	.315	.295	.285	.260	.245	.220	.218
Tip	.215	.190	.180	.150	.135	.110	.090	

No. 15

	End	6″	12″	18″	24″	30″	36″	Ferrule
Butt	.380	.380	.360	.340	.320	.305	.295	.296
Middle Section	.295	.280	.260	.250	.235	.210	.195	.195
Tip	.195	.180	.165	.145	.125	.100	.085	

Has medium dry fly action and weighs 5¾ ounces with solid butt and screw type reel seat.

FLY RODS—SIX-STRIP CONSTRUCTION

3 PIECE

No. 16 8½-Foot. Powerful action.

	End	6″	12″	18″	24″	30″	34″	Ferrule
Butt	.380	.380	.370	.350	.320	.290	.280	.281
Middle Section	.275	.260	.250	.230	.220	.195	.185	.187
Tip	.185	.175	.160	.135	.118	.090	.080	

No. 17 8½-Foot.

	End	6″	12″	18″	24″	30″	34″	Ferrule
Butt	.370	.370	.350	.335	.315	.285	.265	.265
Middle Section	.265	.255	.245	.235	.210	.190	.175	.171
Tip	.170	.158	.148	.125	.103	.080	.070	

A fine, light fly rod, with smooth, easy action, weighing 4¾ ounces with light reel seat.

No. 18 7½-Foot. Very light action.

	End	6″	12″	18″	24″	30″	Ferrule
Butt	.345	.345	.320	.290	.260	.240	.234
Middle Section	.235	.225	.210	.195	.170	.150	.148
Tip	.150	.145	.125	.105	.080	.065	

FLY RODS—FIVE-STRIP CONSTRUCTION

2 PIECE

No. 19 9-Foot. Powerful action.

	End	6″	12″	18″	24″	30″	36″	42″	48″	54″	Ferrule
Butt	.390	.390	.385	.380	.370	.360	.350	.310	.295	.265	.250
Tip	.265	.255	.240	.220	.200	.170	.150	.125	.105	.085	

Weight with slightly hollowed butt and screw type reel seat, 5¾ ounces.

No. 20 9-Foot, having smooth, pleasant action suitable for both wet and dry fly work.

	End	*6″*	*12″*	*18″*	*24″*	*30″*	*36″*	*42″*	*48″*	*54″*	*Ferrule*
Butt	.375	.375	.370	.365	.355	.345	.325	.295	.275	.260	.250
Tip	.260	.245	.230	.205	.190	.160	.140	.120	.100	.085	

Weight with slightly hollowed butt and standard reel seat, 5⅛ ounces.

No. 21 8½-Foot. Light action.

	End	*6″*	*12″*	*18″*	*24″*	*30″*	*36″*	*42″*	*48″*	*51″*	*Ferrule*
Butt	.370	.370	.360	.350	.330	.310	.290	.270	.250	.245	.234
Tip	.240	.225	.210	.190	.160	.140	.120	.105	.085	.070	

Weight, with light reel seat, 4½ ounces.

No. 22 7½-Foot. Powerful dry fly action for light rod.

	End	*6″*	*12″*	*18″*	*24″*	*30″*	*36″*	*42″*	*45″*	*Ferrule*
Butt	.320	.320	.310	.300	.290	.275	.250	.230	.225	.218
Tip	.225	.205	.190	.175	.150	.125	.100	.080	.070	

Weight, with light reel seat, 3½ ounces.

No. 23 7-Foot.

	End	*6″*	*12″*	*18″*	*24″*	*30″*	*36″*	*42″*	*Ferrule*
Butt	.300	.295	.280	.275	.260	.240	.220	.205	.195
Tip	.205	.185	.175	.155	.135	.110	.085	.065	

BAIT CASTING RODS—SIX-STRIP CONSTRUCTION

No. 24 6-Foot. Two-piece, designed for ⅝ ounce lures. Granger Rod.

	End	*6″*	*12″*	*18″*	*24″*	*30″*	*36″*	*Ferrule*
Butt			.325	.295	.280	.250	.235	.234
Tip	.235	.210	.185	.165	.145	.120	.105	

No. 25 By E. C. Powell Rod Co.

	End	*6″*	*12″*	*18″*	*24″*	*30″*	*36″*	*Ferrule*
Butt			.325	.295	.265	.245	.220	.218
Tip	.215	.195	.175	.150	.125	.110	.090	

No. 26 One-piece, for ⅝ ounce lures.

Butt	*6″*	*12″*	*18″*	*24″*	*30″*	*36″*	*42″*	*48″*	*54″*	*60″*	*66″*	*72″*
.380	.380	.350	.300	.270	.240	.215	.195	.180	.150	.130	.110	.100

A favorite rod for accurate tournament work.

BAIT CASTING RODS–FIVE-STRIP

6 FOOT

No. 27 One-piece, tournament action for ⅝ ounce lures.

Butt	*6″*	*12″*	*18″*	*24″*	*30″*	*36″*	*42″*	*48″*	*54″*	*60″*	*63″*
.390	.340	.295	.260	.235	.205	.180	.165	.140	.118	.105	.100

Fitted with ⅜ inch, or .375 ferrule, used with offset handle, total rod length is six feet.

No. 28 Two-piece, for ⅝ ounce lures.

	End	*6″*	*12″*	*18″*	*24″*	*30″*	*36″*	*Ferrule*
Butt	.360	.350	.330	.315	.275	.260	.245	.234
Tip	.245	.220	.200	.180	.150	.130	.105	

SPINNING RODS

2 PIECE

No. 29 Five-strip, length 7 feet, 5 inches. By Uslan Rod Company.

	End	*6″*	*12″*	*18″*	*24″*	*30″*	*36″*	*42″*	*44½″*	*Ferrule*
Butt			.305	.305	.295	.275	.250	.230	.225	.203
Tip	.215	.210	.200	.190	.175	.148	.125	.100	.095	

No. 30 Six-strip, length 7 feet. Designed by Robert Crompton.

	End	*6″*	*12″*	*18″*	*24″*	*30″*	*36″*	*42″*	*Ferrule*
Butt			.266	.262	.254	.242	.216	.188	.187
Tip	.188	.184	.180	.162	.148	.130	.120	.110	

SALT WATER RODS

No. 31 Tuna Club "3–6" Rod. By South Bend Company.

End	*6"*	*12"*	*18"*	*24"*	*30"*	*36"*	*42"*	*48"*	*54"*	*61½"*
.435	.410	.385	.370	.350	.315	.285	.250	.215	.180	.155

Fitted with male ferrule, five guides and tip top, this rod will weigh slightly under four ounces.

No. 32 Tuna Club "Light Tackle" Rod. By South Bend Company.

End	*6"*	*12"*	*18"*	*24"*	*30"*	*36"*	*42"*	*48"*	*54"*	*61½"*
.560	.530	.500	.460	.440	.385	.350	.310	.270	.225	.200

Completely fitted, this rod weighs 6 ounces.
The above rod is double-built, and only the best of thick walled cane will serve if single-built construction is attempted.

CHAPTER 5

BUILDING YOUR OWN ROD—MAKING READY

Trepidation may assail you here, for the task will seem tremendous after the brief foregoing description of rod construction in general, in which I have tried to emphasize the necessity of painstaking and accurate work to produce a fine rod. With the decision made to start, be determined to carry through, with my assurance that the many potential difficulties which now seem so confusing will become magically ironed out once you are well into the work.

You will find the various steps of the work interesting, the keen "snick" of your plane as it removes a long, thin shaving from the spline, music to the ear. The use of the micrometer in determining exact calibrations becomes a pleasure. The completion of that first set of splines, exactly fitted and ready for the glue, will prove an event which will live long in your memory. And finally that completed rod, your *first rod,* a live, vibrant, beautifully finished instrument fashioned with painstaking care by your own hands, will become your most valued possession. Later rods, which you will already be planning, will show improved craftsmanship, have better action and balance, perhaps, but never will they replace that first rod in your affections.

First, the bamboo should be ordered. The canes usually come 8 feet long and from 1½ to 2 inches in diameter. One such cane will provide material for about two fly rods, or three shorter rods, so it is wise to order several of the best grade obtainable, since the cost is nominal. This ample supply of stock will permit careful sorting of the best strips after splitting, and additional seasoning of the remainder for later use.

I have had very fine bamboo from the importer, Charles H. Demarest Company, Water and Bleekman Streets, New York City. Also supplying good stock in a variety of grades is Herter's Supply House, Waseca, Minnesota. This firm also handles an excellent steel planing form, suitable for the amateur workman, and the marvelous little pressure winding machine described in Chapter 11. If, after consideration of the various planing forms described in the next chapter, you wish to purchase the Herter form, it should be ordered along with the winding machine.

There are, no doubt, other importing houses who could supply good bamboo in small quantities. And the large rod manufacturers would at one time sell canes, but it is doubtful if now they would let go any of their choice, well-seasoned stock. With the unsettled conditions in southern China at present there may possibly be difficulty in bamboo shipments for a long time to come. This is another reason why your initial order of bamboo should insure enough stock for at least several rods.

SMALL TOOLS NEEDED: (Minimum)

One carpenter's plane, 8 or 9-inch, of the best grade obtainable.

One "low angle" plane, 6-inch, for fine finishing cuts. Both of these in the Stanley make are excellent.

Heavy-blade hunting or butcher knife.

Light mallet.

12-inch mill file. The Heller "Nu-cut" probably best.

Small triangular hacksaw.

Half-round gouge chisel, for removing node "dams."

One-inch micrometer.

Outside caliper, for gauging cork grips, etc.

Small breast drill. Drills size .031 and .038, for fitting pins in ferrules and reel seats.

8-inch fine-grain oil stone of the "Carborundum" type, for first rough edge on plane blades.

6-inch oil stone of the finest grit, for finish edge on plane blades. The "Lily White Washita" stone, by the Norton Abrasive Company is very fine.

Machinist's 60-degree center gauge. For final finishing splines for six-strip rods.

A good bench vise and the usual small tools possessed by every sportsman will complete the necessary implements needed for nearly all rod work. A small bench lathe will prove of inestimable value in several phases of the work and if you do not own one perhaps a friend will oblige one evening, when your work has progressed toward the finishing stages.

With your tools assembled, and while waiting for your bamboo, planing form and winding machine, there will be opportunity to study rods of various types and calibres, and decide what your first handmade rods shall be. Perhaps you have a favorite rod you wish to duplicate, or have long admired a fine rod belonging to a friend. You may very well have either. And it is far safer to build your first rod to match one of proven tapers than to experiment with improvements. Since slight changes in calibrations will affect rod action markedly, the fun of this experimenting can come later as your handicraft improves and experience tells you just what rod action will be attained by a variation in tapers. By then, no doubt, you will have a truly revolutionary design worked out in your mind, perhaps a rod for a special purpose, entirely different from anything to be found on the market. And the building of this, your own "brain child," will prove high adventure, with long hours devoted to casting and flexing tests of various rods and the final charting of the new one as your ideas crystallize into micrometer readings to which you will work.

If a fly rod is to be your first attempt, do not fear that small tip which, when divided into its tiny strips, may at first seem impossible to construct. Actually, small strips are easier to plane and fit than those for butt sections of a rod. The reasons will appear as you progress with the work, and are explained in Chapters 8 and 9.

But perhaps the easiest rod for your initial attempt would be a bait casting rod, consisting of butt and tip, with one ferrule. In this, the butt section will not be so heavy as to offer difficulty in the planing, while the tip will require less care in shaping than would the smaller sections of a fly rod tip. So if you wish to copy a favorite stick of this type, lay it out on your work bench beside a yard stick and use the micrometer to take over-all readings, across the "flats" at 6-inch intervals, from butt to tip. Various proven rod charts are included in Chapter 4, any of which might be used, but assuming that you wish first to copy a rod, the chart should be like this:

SIX-FOOT BAIT CASTING ROD
(Light action)

Two-piece. Ferrule: 7/32 inch, or .218.
Micrometer readings at each six-inch station:

Butt section:

Butt	*6″*	*12″*	*18″*	*24″*	*30″*	*36″*	*Female ferrule*
.375	.375	.330	.295	.265	.240	.220	.218

Tip section:

Butt	*6″*	*12″*	*18″*	*24″*	*30″*	*36″*	*Tip-top 6/64″ or*
.220	.190	.170	.150	.130	.115	.100	.093

If you are not familiar with the micrometer, this preliminary work will be of practical value in learning to read quickly that little instrument whose accuracy will guide you through all the succeeding work. Note that on the sleeve there are ten divisions, marked from 1 to 10. With the micrometer fully opened until the zero on the thimble is opposite the figure 10, the reading would be exactly one inch. Thus, the zero stopped at figure 1 would give a reading of one tenth of an inch or .100. Now there are 25 lines around the thimble, each representing one thousandth of an inch. Thus, to "mike" a rod from the above chart at, say, the 18-inch station for the butt joint, the thimble is revolved until the zero is at figure 3 on the sleeve, giving us a reading of .300. Then turning the thimble back 5 divisions to the figure 20, gives us five thousandths less than .300, or the desired reading of .295.

Now, too, will be a good time to become familiar with your planes, for these will be in constant use later in all the accurate work of shaping the splines. The time will come when their handling will be almost automatic, but now practice taking out the irons—or blades—and resetting them to take the very finest possible cut. Then work on a narrow strip of soft wood, and try to take a thin shaving the full length without lifting the plane. The knack of this will soon come to you and will later be of value in the bamboo work.

This is also the time to practice planing a strip of wood, until the plane can be held level with the work without conscious thought. At first you must follow its travel with the eye, correcting any tendency to cant it to either side. Later, with sufficient practice, a sort of "sixth sense" will seem to develop, after which it is actually an effort to tilt the tool. It is difficult to explain just how this occurs, but the pleasing result is something akin to the act of pointing a finger at an object—wholly automatic.

All this preliminary practice, while tiresome now perhaps, although the use of a fine edged tool is always pleasurable, will pay ample dividends later in the bamboo planing. Then, when the splines are being shaped to the true triangle that is so essential, the accurate holding of the plane—always level with the face of the planing form—is the sole gauge which will insure accurate work.

SHARPENING THE PLANE IRON.

Most of us can sharpen tools, and quite adequately, too, for ordinary use, but the edge for bamboo work is something else again; it must be actually razor-sharp, as keen as the structure of the steel in that blade will permit. We will hope your two planes possess high quality steel in their irons; and all the best grade Stanley planes I have used have had good irons, properly tempered.

The regular angle to which the irons are ground will be proper for a good edge, and one much sharper than if they are ground off to a more acute angle, sometimes advised for bamboo planing. While a blade so ground does cut with somewhat of a "scraper" effect, and so does not readily tear the bamboo at the interlaced grain of the nodes, I have found it far better to work with the longer angle, and trust to an ultra-keen edge for smooth work over the nodes, along with other little tricks covered fully in Chapter 8.

Using the fine side of your 8-inch oil stone, first treat it with a plentiful sprinkling of oil made of No. 10 lubricating oil and kerosene, in equal parts. This used on both your stones not only makes them cut faster, but floats off the tiny particles of steel which come from the tool, and which would otherwise become imbedded in the stone.

The iron should be held firmly but easily in both hands, with a sort of overlapping grip, and passed back and forth the full length of the stone. Being careful to hold the angle of the cutting edge flat at all times, will obviate the necessity of soon regrinding it. A slight tipping up from the proper angle will give a quicker edge of course, and there is always a temptation to do this, but the time gained will later be sacrificed in the laborious job of entirely refacing the plane iron to the original angle.

Now, when is the tool sharp? The common test is to run a finger lightly along the edge; but this will not be adequate. Instead, rub the ball of the finger flat along the back of the iron, and *feel* for an infinitesimal wire edge as it passes over the end of the blade. The eye cannot detect this tiny turned up sliver of steel, but its presence the full width of the blade indicates you will have a real edge. But this is not all. Lay the blade flat on the stone and rub a few times to turn the wire edge back flush with the angle of the blade face. Now it is loosened—that thin hair of steel which will come off when the edge is passed strongly across a piece of hardwood.

A *rough* sharp edge will remain, good enough for all wood planing, but still not keen enough for bamboo. This final keenness is obtained on the "Lily White Washita" stone, which apparently has no grain at all when you pass a finger over it, but which will turn down another very fine hair of steel when you proceed with a few passes as was done on the first stone. Remove this on the wood block, and now your plane is really *sharp*. If of good, fine-grained steel it should take the hair from one's wrist like a razor.

I trust that the beginner in this kind of work will not feel overwhelmed after reading the above detailed description of the one really simple operation of plane sharpening. But this knowledge will prove helpful later. Then, when really into the bamboo work, the necessity of a keen blade will be so evident that renewing it will become a delightful job. Actually, a blade can be newly keened in about three minutes, and the resulting pleasant "snick" as it enters the hard bamboo and turns up a shaving thinner than paper, the full length of a spline, will prove highly gratifying.

OTHER PRACTICE JOBS.

The casual "rod tinkerer" who may have had some experience in rod overhauling, refitting ferrules and the like, may already feel competent to finish his carefully made rod sections when the time comes. But the new hand at the game can well practice a bit. Proper fitting of ferrules alone, on those rod sections over which you will labor so carefully, is something of a trick, requiring extreme care.

Thus, you might remove a ferrule from an old, discarded rod, saw off the wood, form a new seat, as advised in Chapter 13, and reset the ferrule. And the complete refinishing of a favorite rod, using the methods later described for "new work," will prove useful in learning the technique of wrapping and varnishing.

CHAPTER 6

PLANING FORMS

The planing form, sometimes referred to as a "V" block, with its accurately milled grooves, will hold the bamboo strips for planing and also serve as a rough gauge as the work progresses to the point where the micrometer is used. There are two types of steel forms available for the amateur workman, both capable of producing extremely accurate work, and each possessing special advantages.

So accurate, inexpensive and easy to procure is the steel form, there is no longer the necessity of laboriously building, or having constructed, a wood form as was necessary a few years ago.

THE ONE-PIECE STEEL FORM.

Happily, the Herter Company, referred to in the last chapter, has anticipated the needs of the amateur rod builder, since this work has become so popular among anglers, and is supplying excellent specialized equipment. I know of no other firm in this field.

Their planing form is a slab of mild steel, 41 inches in length, about 5 inches wide and ¾ inch thick. The six-strip form has four parallel 60-degree grooves, each carrying its length with a constant taper reduction of .008 inch for each six inches, except at the larger end of each groove, where a slight lessening, or overlap, permits easier planing as the bamboo strip is changed from one groove to the next.

The largest groove starts with a depth of .2261 inch, sufficient for rod butts up to the largest fly and bait casting rods, and even for the lighter salt water rods. The small end of the last groove

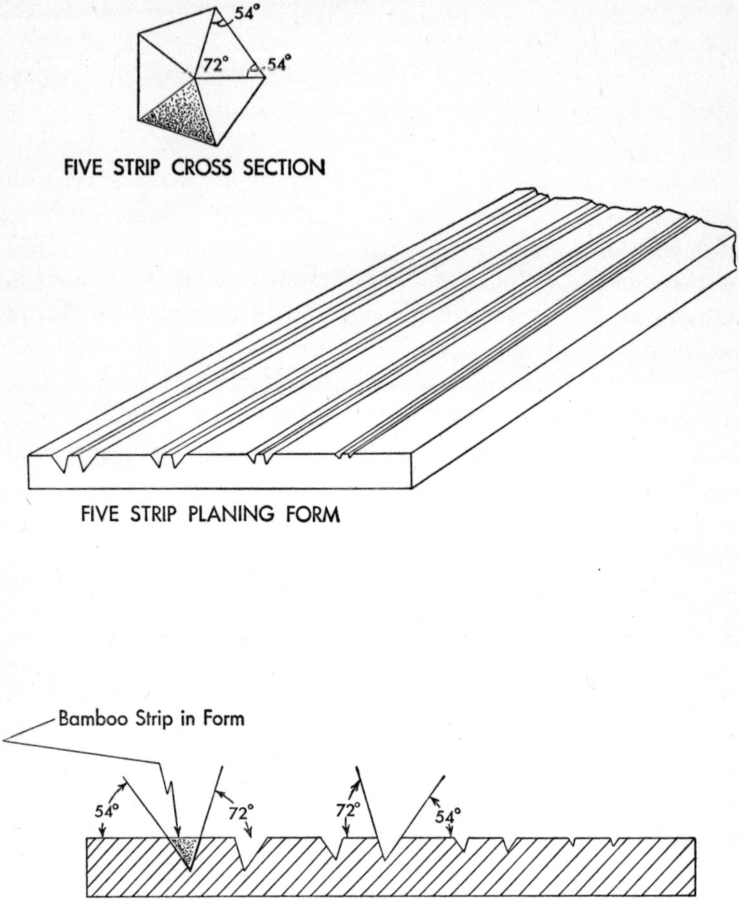

Five-strip planing form.

has a depth of only .0312 inch, thus permitting a small fly rod tip only 4/64 inch in diameter.

A similar five-strip form is also offered by this company, with grooves of similar depth and taper. In five-strip construction, however, the rod splines are not an equilateral triangle as are those for the six-strip rod. The grooves must be cut to produce a 72- and a 54-degree angle on each spline (note illustration). Thus, there will be four pairs of matching grooves, the bamboo strip being turned from one of a pair to the next as planing progresses.

Either of these forms will cost only about $25, or perhaps a bit more with transportation costs added. A combination form, also offered, is milled for six-strip work on one side and has five-strip grooves on the other. It will prove well worth the cost, something like $35, for the serious rod builder who wishes to work with both types of rods.

THE UNIVERSAL TWO-BAR STEEL FORM.

This type of form is radically different from the one above, and will require "setting" for rods of different sizes. This feature, however, since the form is easily adjustable and 48 inches in length, will turn out six-strip rods up to a butt diameter of almost ¾ inch, and down to the smallest tip size of 4⁄64 inch.

This form is not for sale, so far as I know, but the bars of mild steel, ¾ inch in diameter, can be provided by and milled in any good machine shop. The square bars should first be milled lightly on all sides to prevent warping and provide a smooth surface for your later work. Then have matching corners of each bar beveled to an angle which will provide an exact 60-degree triangle when the two are placed together. Note that these grooves will not taper, but will be of constant depth.

Specify as follows:

The largest groove, depth: .312 inch.

Second groove, depth: .218 inch.

Third groove, depth: .109 inch.

Fourth groove, depth: .031 inch.

For mounting the bars, after milling, procure a billet of well-seasoned maple or other hardwood, smoothly planed, about 2½ by 3½ inches in size and 50 inches long. Be sure that one narrow edge of this is perfectly true, and then chisel inlets for five, 3-

inch "C" clamps, to be mounted a foot apart, as shown in the illustration. These should be tilted so their jaws will clear the upper surface of the bars by at least ¼ inch. The body of each clamp should be drilled for two wood screws to hold them securely, after the set is carefully lined up.

Now, with the bars snugly clamped in place, the groove depth will, of course, be constant. To secure any taper desired, thin shims are inserted between the bars at each clamp and halfway between the clamps. This will permit the same method of gauging as with the flat form and match the six-inch intervals at which calibrations are shown on your rod chart.

Shims may be cut from brass, tin or aluminum, and should run from a maximum thickness of about .025 down to .002 inch, with various graduations which will permit any desired thickness. A size about ½ inch square will drop between the bars, to rest upon the wood block and clear the bottom of the "V" in which rests the strip.

The chart to be used for a rod made in this form may be made up exactly like those listed in Chapter 4, but with the shim thicknesses for each 6-inch station calculated in advance, and added to the chart. For example, a fly rod tip made in the smallest groove can be charted in this manner:

	End	*6″*	*12″*	*18″*	*24″*	*30″*	*36″*
Tip diameter	.203	.180	.170	.150	.120	.090	.062
Groove depth	.031–Full length–						
Shim thickness	.071	.059	.054	.044	.029	.014	.000

Thus, for example, starting with the large end of the tip, we find that the groove depth, .031, plus the shim thickness, .071, will result in a spline measuring .102 from surface to inner apex after it is planed flush. And, since this spline measurement is always just half that of the total rod diameter in the six-strip, the matter becomes simple, the completed rod stick finishing out within one or two thousandths of an inch of the calculations.

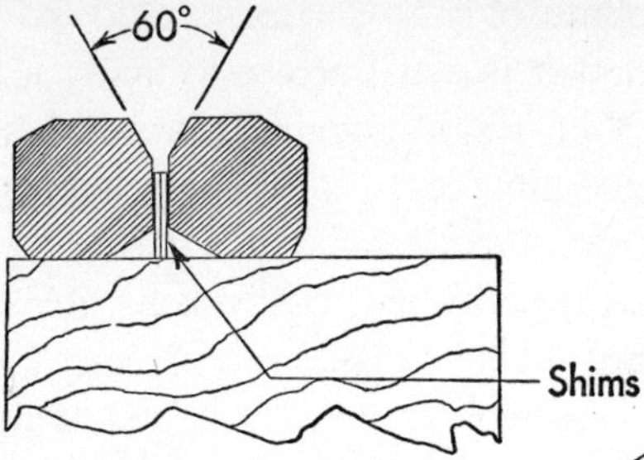

CROSS SECTION ADJUSTABLE FORM

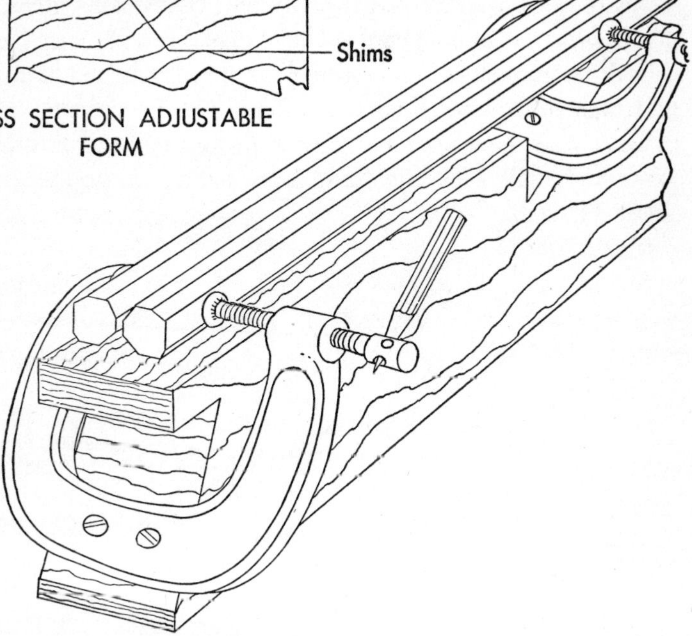

ADJUSTABLE STEEL PLANING FORM

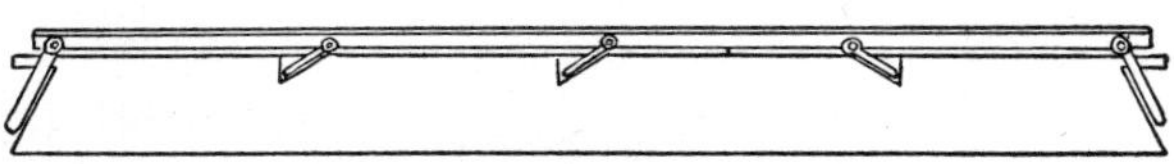

SIDE VIEW – ADJUSTABLE FORMS MOUNTED ON MAPLE BLOCK

The adjustable steel bar planing form, suitable for building rods of any size, up to light salt-water rods.

In actual practice the splines, of course, cannot be planed down quite flush to the steel, and it is usually best to allow a slight tolerance by using shims of slightly less thickness than the calculations prove necessary, thus leaving a bit of clearance for the plane blade. In any case, understand that the planing form serves only as a tentative gauge. Actual calibrations are always taken with the micrometer, as later explained.

This assembled form is best held on the work bench by drilling a ¼ inch hole in the rear end—from which you plane—another in your work bench, and inserting a short steel or brass dowel pin which will keep the form from sliding or tipping and is easily removable.

CHAPTER 7

STARTING THE WORK

At last, with your bamboo on hand and your planing form ready, you will start on your first rod. Long before this eventful moment you will have decided on the rod to be attempted, with a chart prepared. It is best to ink or type this on a card to be tacked on the wall back of your work bench for ready reference.

The bench, over which you will labor so many delightful hours, should have sufficient height to save tiresome stooping when you are using the plane. A length of 6 or 8 feet will be ample, and the width is immaterial, but its top should be at least 36 inches from the floor if you are of average height.

INSPECTING AND SPLITTING THE CANES.

Your canes, probably 8 feet in length, should be fairly straight and, if well seasoned, of a uniformly tan or yellow color. Any of those showing a green tinge should be exposed to the sun for additional curing. And those not for immediate use may well be stored in a warm, dry place, even when apparently well seasoned, to prevent the absorption of moisture from the air, if you live in a damp climate. But first split these in halves—marked for later matching in pairs—to prevent checking, which will sometimes occur in the whole cane if these are left for a long period of final seasoning before you are ready to work them up.

It is best always to use strips from the same cane for one rod section, and for an entire rod when possible, since each cane will vary in its structure, and mixed strips may possibly set up opposing stresses in the rod, causing a slight warping of the stick, perhaps, and unreliable action. By "putting back" the strips into the rod, much as nature grew them, a more harmonious performance of the whole will be better assured.

Halve the canes by starting your heavy knife at the butt with a sharp tap of the mallet. Force the blade with a twisting motion down to the first node. Use another tap to break through the tough "dam" there, and so on through. The halves will separate neatly and easily. Next lay each half on your bench and knock out each half dam at the nodes with the half-round chisel

Before going farther inspect the end grain of each cane at the butt, and choose a half showing enough good dense fibre

Splitting the cane and separating the strips.

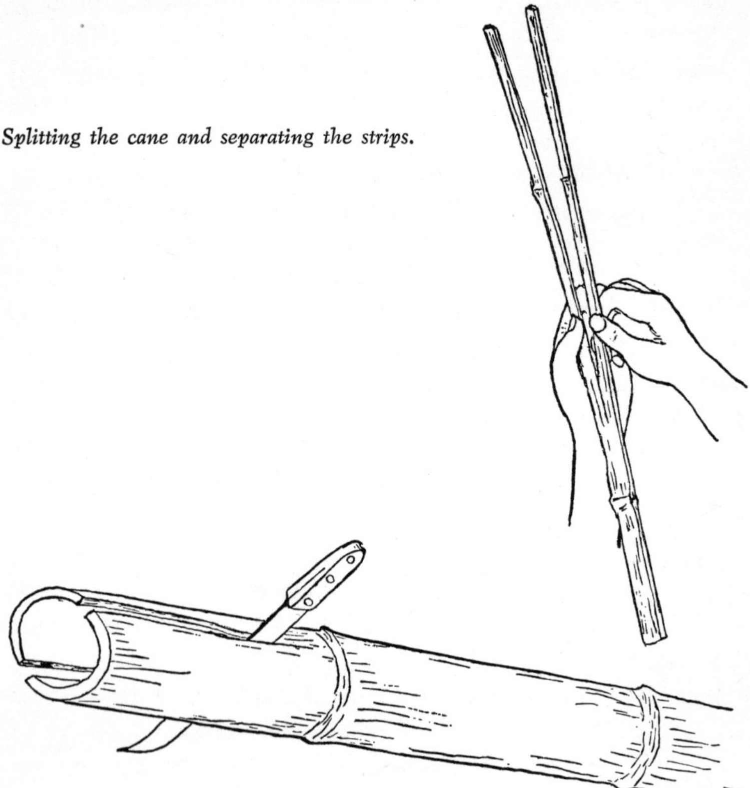

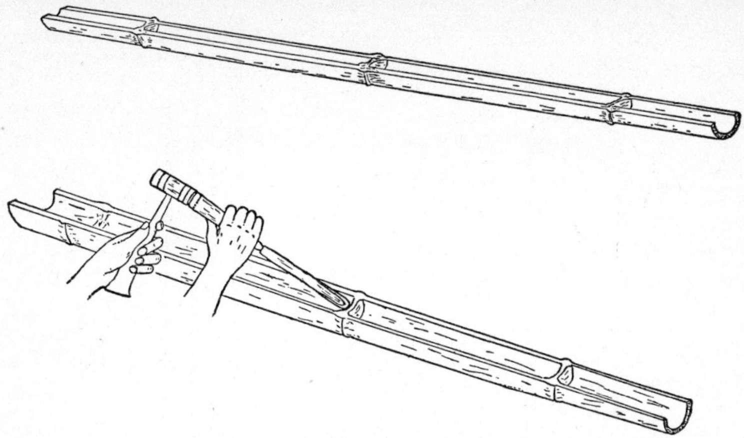

Halved cane and method of removing node "dam" with half-round chisel.

to extend in to the center of the rod you will build. Immediately under the surface enamel will be a narrow area of dark brown fibres, so fine as to appear a solid mass. These are the real *power fibres,* which will impart to the finished rod its smooth, resilient action and strength. A fly rod tip and other small-diameter rod sections will contain this best material in to the core. Butt sections may perhaps contain a little of the less dense area which occurs between these outer fibres and the interior pith, but without harm, since the over-all diameter of the butt is considerable and it does not undergo the same flexing strain as the smaller portions of the rod.

Let us assume that you will start a 9-foot, three-piece fly rod, the sections of which will be 36 inches long. With the little hack saw, cut the cane in half, giving you a length of 48 inches. The butt section will split out six or eight strips, amply wide for the rod butt, leaving an extra strip or two in case one of your strips should later be ruined or show a bad spot.

Now halve the half cane and split these resulting quarters again into three or four strips each. Tentative pencil marks at

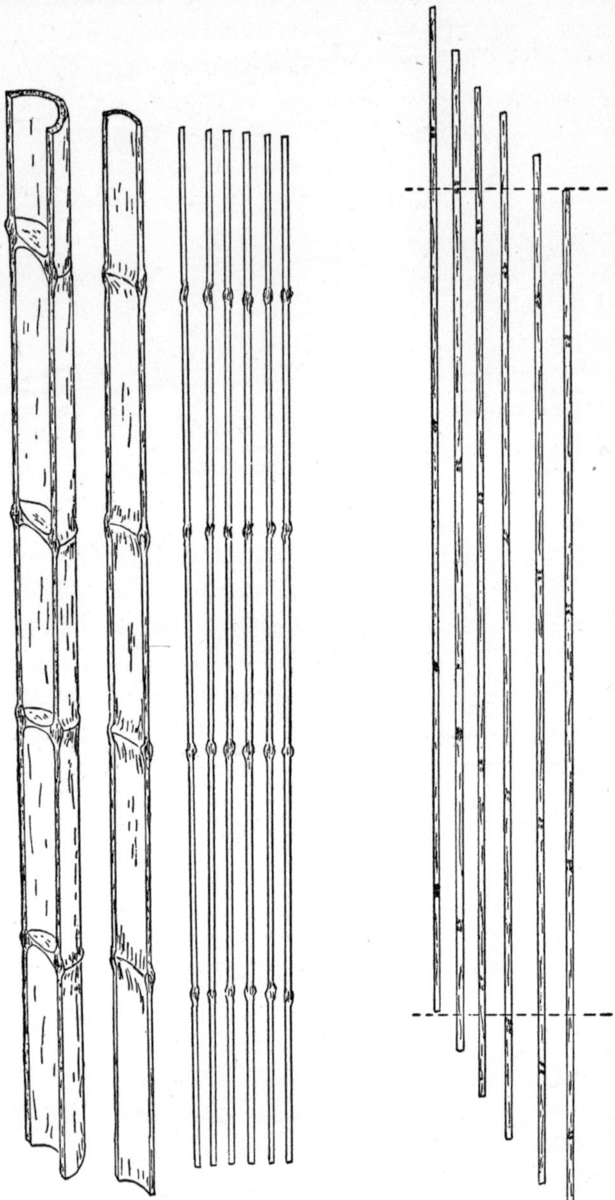

1 2 3

the butt end before the knife is started will indicate the possible number of strips, each of which should be at least $\frac{1}{16}$ inch wider than the face of the finished spline is to be. Splitting of the quarter canes is best done by starting the knife at the center each time, then grasping the opened ends and rending them apart with the hands. A bit of practice will enable you to vary the pressure, with either hand, to correct any tendency the split may have to run crookedly.

Even the best of bamboo may sometimes contain a bad spot which should be eliminated. So test each strip by grasping it between the hands and flexing it to a considerable arc, "feeling" for any suggestion of a soft or "doughy" spot. If present, this may be due to a bruise on the cane surface or to an unusually deep hollow next to a node. Ordinarily the flexed strip will spring back smartly to its original shape. If it does not, discard the strip and substitute another from your extra strips.

Now mark off two lines on your bench top, 37 inches apart, lay the 48-inch strips side by side over the marks and slide each past its neighbor an inch or more, thus separating the nodes and so evenly distributing the weak spots at these points where the grain is crossed and interlaced. Saw off the projecting ends of the strips at each mark, then number each from 1 to 6 with a pencil mark on the exterior surface. The strips will later be glued up in that order, thus spiraling the nodes evenly in the finished stick.

The remaining butt half of this cane will provide enough

◀ METHOD OF STAGGERING THE NODES FOR A ROD SECTION

1 *Halved cane before removing node "dams," and quarter section with these removed.*
2 *Strips split out for rod section.*
3 *Strips moved to space nodes equally, and ends cut off at dotted line.*

strips for the rod's middle section, with some to spare, insuring fine, dense fibre in to the core. And the upper 48 inches of a half section will usually split out a dozen or more narrow strips for the two tips. Use the same technique as for the heavier rod sections in splitting and cutting these, and give each strip a severe flexing test; tip strips should be cut 37 inches long.

It is best to make ready the entire set of strips for a complete rod before proceeding farther. Then mark each stick at the butt end with a dab of ink, to prevent reversing it during the succeeding operations. Each set of six should, of course, be bundled up separately.

PREPARING STRIPS FOR A TWO-PIECE ROD.

While the conventional fly rod is usually of three sections, you may prefer two-piece construction, which has several advantages and is rapidly becoming very popular. The three-piece rod has two sets of ferrules, thus two flat spots which disturb its even arc in use. The two-piecer, with its one ferrule at the center has only one interruption of its smooth arc, and that at the center, permitting free play of the important upper half of the stick.

The one disadvantage in the two-piece rod is perhaps more difficult transportation, but certainly a rod up to 9 feet in length, having 54-inch sections, will not prove unduly clumsy to carry in its tubular case.

For a rod butt of this type, cut the butt section of your half cane at least 60 inches long, and trim the split strips between lines 54½ inches apart on your bench. The tip strips should be cut 55 inches long, since the extra inch will be "used" by the male ferrule when the rod is completed. These will be made from the second half of your cane, the 60-inch section being measured down from the tip end. In thus using the upper cane for tips there will be fewer nodes in each strip, since these occur more closely in the lower section, the intervals always widening toward the upper end.

If a two-piece rod is attempted, it will be found to require less hours of work than a similar rod in three-piece construction, since there will be six strips fewer to handle all through the various operations. Planing the longer strips may at first be a bit more difficult, and the gluing may require more care, but in the essential phases the work is the same.

SQUARING UP THE STRIPS.

Your split strips will all have rough edges and be more or less warped, a matter that will be disturbing for the moment, but which will be taken care of later. The exterior collar of the nodes should first be removed and leveled with the surface with your big file. A power sandpaper wheel, if available, will do this work quickly and neatly.

Now plane the strips roughly square, by holding each on the bench and using your large plane on the pith side and two edges. *Never,* at any time, during these operations, disturb the outside surface enamel. Wear a glove on the hand which holds down the strip, and shift it from ahead to behind the plane as each stroke is made.

This simple technique of "holding down," which soon becomes automatic, will also be used for all work in the planing form later. The plane should be set to take a moderately light cut for this work, and can be moved rapidly in short, overlapping strokes. The aim now is only to take the rough edges and inside pith from the strips.

HEAT STRAIGHTENING THE STRIPS.

Bamboo, fortunately, has the faculty of becoming soft and pliable when warm and at once returning to its original state when cool. This renders the straightening of the more or less warped strips a simple process. There may be a few minor, short kinks at the leaf nodes and they often will be mildly convex, bowing out from the enamel side.

Wear gloves and pass the strip over a gas burner or across the top of a stove, moving it continually back and forth, until the wood is quite warm. Then quickly sight down its length and counter bend as required, reheating and repeating the process until the strip is perfectly straight. After each strip has cooled, examine it again, with particular attention to the enamel side. Only if this side lies straight and true in the grooves of your planing form can you complete a spline with the accurate angles necessary for perfect gluing.

Your strips should now lie straight as arrows, side by side on your bench, requiring perhaps only a bit more leveling at the nodes to be ready for beveling.

CHAPTER 8

SIX-STRIP RODS—PLANING AND FITTING

Now you will have reached the first part of the real precision work and will be anxious to complete your first rod section. If you use the flat type planing form, first cross mark its surface with a light file cut at each 6-inch station, for easier reference when using the micrometer in connection with your chart.

Have the large plane very keen and set to take a medium cut at first. It is best to start with the largest rod section, laying the strip in the largest groove of the form, with the enamel side *always* against the side of the groove. Thus, one of the inner corners to be planed away will always be uppermost (note Fig. 1 in the illustration). Figs. 2 and 3 show how this reduction gradually forms the inner apex of the spline.

Use short, overlapping plane strokes to take off a bit of the first corner, turn the strip with the enamel resting in the other side of the groove, and repeat. As the planed surfaces widen you may have to adjust the plane for a finer cut, to prevent tearing the bamboo at the nodes. And turn the strip about every second cut, to reduce both sides equally. This turning also guides the accuracy of your work, since the beveled side fitting into the groove requires that the plane then need only be held level with the face of the form.

Lay aside each rough-planed strip when the inner apex is attained, to be finished with the smaller plane, and well before it approaches the final dimensions, as determined by trial calibrations at each station. It is best to complete a set of splines in this manner, before the final, more accurate planing. Splines for smaller rods, of course, will be transferred from one groove to

1

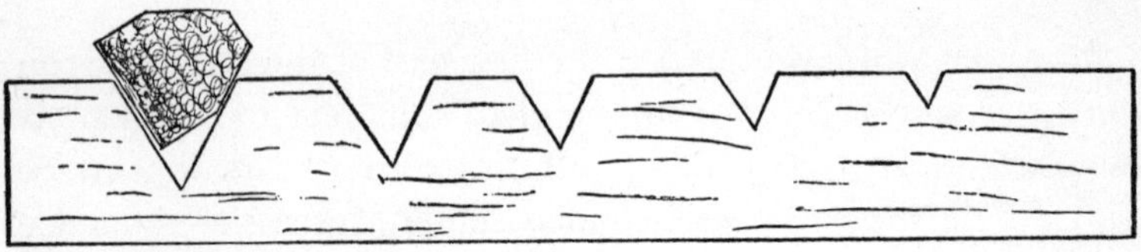

2

3

Cross section of six-strip planing form, showing how squared strip is reduced to final triangular shape. Enamel side of strip always rests in groove.

the next smaller as the corners are reduced and the spline lowers in the groove. And you will find it handy at times to slide the strip forward and back, to utilize the largest possible groove, with still enough wood projecting above the metal so that a cut can be made without nicking the form.

The longer strips for two-piece rods will necessarily be slid in the grooves during the shaping, a procedure which becomes almost automatic, the one gloved hand moving the strip and turning it between every two or three plane strokes.

The 6-inch low angle plane has the blade set much lower than the standard tool, with its bevel turned upward instead of down, thus permitting the removal of a very fine shaving, which is highly essential in the final shaping work. Ordinarily, we push a plane straight with the work, but an easier cut is possible with

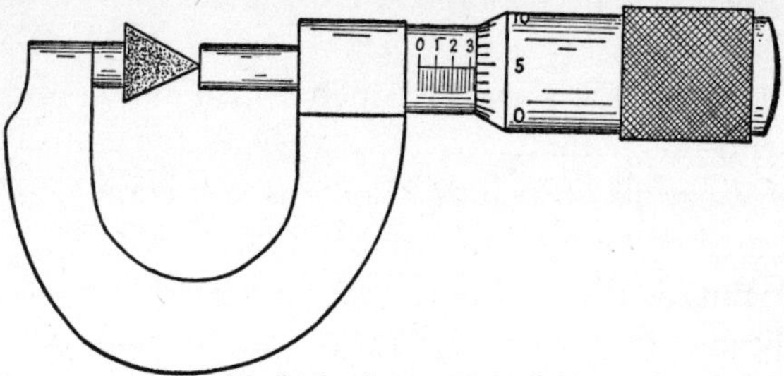

Method of gauging splines with micrometer, used in both five- and six-strip work.

the tool turned to about a 15-degree angle, which takes off the shaving with a "shearing" cut, with less danger of tearing the wood at the cross-grained nodes. The blade at all times must, of course, be kept razor sharp.

Turn the strips after each light cut with the little plane, and stop often to calibrate them at each station. Remember that the thickness of the enamel, while slight, will add about .003 inch to the measurements. The finished rod will finally have this thin coating removed, so it is best to add this amount to each calibration called for in the chart, when measuring the stick from apex to flat. While this small difference might be forgotten in the rod butt, you are determined to strive for real accuracy throughout, and there can be no such tolerance in the smaller rod sections.

Thus, when you "mike" the strips for rod tips it is best to take a reading over the enamel at one point, then scrape it off and again apply the micrometer, to determine the exact thickness of the coating on this particular cane. Often it will be no more than .002 inch on the upper part used for rod tips.

Before reaching this point in your work you may be wondering how to secure a slower or quicker taper in the splines, as called for in your chart, when all the grooves in the form are of constant taper. This is attained by sliding the stick forward or back when you reach the portion requiring a change of taper. A little practice will enable starting the little plane at any point where reduction is necessary, and removing a fine shaving without the least noticeable step. A gentle start of the blade will take a shaving no thicker than .002 inch which starts—and can be made to end—at "nothing."

Do not worry about such tiny faults as may be present marring the smoothed surfaces of the splines. These will later be easily removed. But, as the final light cuts are made, and the micrometer indicates that each spline is almost finished, *do* pay special attention to the "always accursed" nodes. Sharpen the plane blade again on your fine stone, set it for the thinnest possible cut, and cross the nodes at an angle. If they are still inclined to tear, file across the offending ones with your big file, so that the next, and perhaps final, passage of the plane will just clear them.

Sometimes a node will cut cleanly in one direction and tear when reversed; so it may be wise at times to turn the strip and form and make the final cut from the small end. Ordinarily, of course, all planing is done from the butt end of the splines.

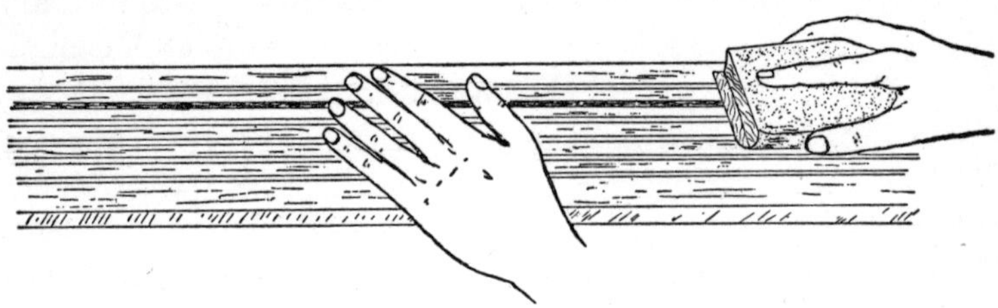

Using sandpaper block to finish planed rod splines held flush in groove of either five- or six-strip planing form.

With careful work, your first set of splines should finally calibrate within .001 or .002 inch of the charted figures at each station. The final treatment now, before fitting them together for inspection, is a light sanding of each planed surface. Prepare a sandpaper block by tacking No. 00 "garnet" paper on its face, lay each spline almost flush in its groove, with its surface so nearly level with the steel that the block cannot be tilted and so change the bevel, and pass the block firmly along its length.

This will remove any slight "step" which may have been left by the plane and smooth any roughness at the nodes. The final operation now, before trial fitting, is to remove the sharp corners of the splines by drawing each over the sandpaper block while the gloved fingers apply pressure on the flat side. Note illustration.

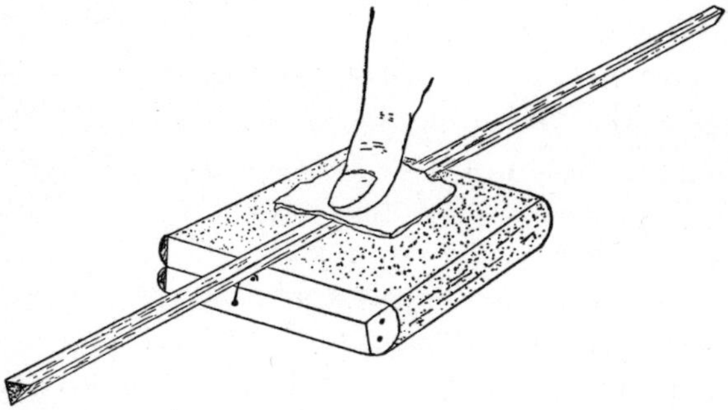

Removing sharp edges of finished spline by drawing across sandpaper block.

Taking off this "hair" from each sharp corner permits closer fitting of the inner apexes, and removes any possible tiny loosened fibre from the outer corners which might split off wider during trial wrapping and gluing operations. Further reduction of the inner apexes of the larger rod sections to secure hollow-built or semi-hollow construction will be explained presently, but now we are concerned with the first proving of your initial efforts with the plane and micrometer.

INSPECTION AND FINAL FITTING.

Have a ball of common grocer's cotton wrapping string, which you will use now, and later for the gluing operations. Fit the splines together, butt ends flush and in the order marked, to spiral the nodes evenly around the stick. The ball of cord may be placed in a can on the floor, to feed off easily, after you make a tight, double half-hitch around the butt of the stick and start revolving it, to secure tight, even spirals its full length. Run the cord under the foot, to secure the necessary pressure, grasp the stick tightly and hold at an angle to space the wraps about a half inch apart. Tie off at the end with a half hitch, and the stick is ready for inspection and calibrating.

Now if every apex of the stick appears perfect, without the tiniest opening, and the entire stick appears to be of solid, unjoined wood, you should pause for a period of well-earned self-congratulation. You will have proven yourself an excellent and meticulous workman. But do not feel badly if a tiny space appears along an edge. Just mark this bad spot with crayon, dissemble the stick, lay the offending spline in its groove, and apply the sandpaper block at the faulty place.

Reassemble the splines, inspect again, and persist in correction until all join perfectly. Remember that a really good rod can result only from a perfect glue job, which in return requires that the splines fit perfectly at all points. Then, too, your pride of workmanship will insist on faultless fitting in this, your first, long-dreamed-of handmade bamboo rod.

A long separation between two splines may indicate that one is not a true triangle—that the plane was not held true with the face of the planing form in making the final cuts. There is a cure for this trouble if the fault is not too great. Your little machinist's center gauge has three 60-degree "V's" of different sizes, so clamp the instrument in a vise, with one V projecting above the jaws. Now, with your gloved hand pressing hard upon the offending spline, draw it through the notch, full length. A tiny shaving will be scraped from the inaccurate surface, and several trips through should entirely remedy the difficulty.

When a final wrapping proves the stick accurately fitted, you may then take micrometer measurements across the flats at each 6-inch station for comparison with the required calibrations on the chart. If your allowance for enamel thickness has been about right, the stick should measure perhaps from .004 to .006 inch over the required chart readings at each point. This extra will be removed when the enamel is finally scraped off and after gluing and seasoning.

To test the comparative accuracy of each spline, "mike" across each pair of flats at each station and compare the three readings obtained. A variation of .002 or .003 inch will be quite permissible, I can assure you, and will result in a rod having an even flow of action, without a "flat spot," regardless of which way it

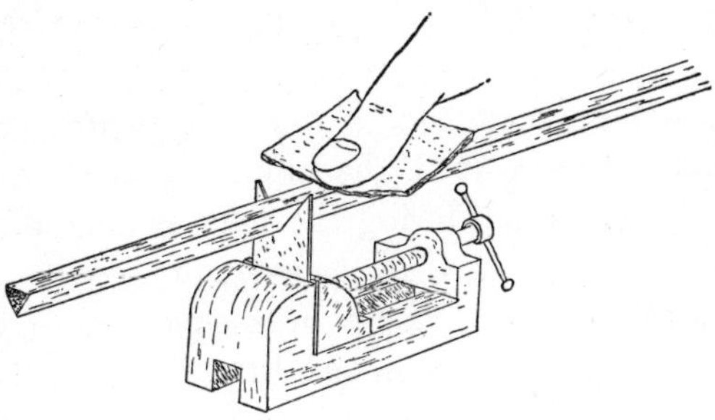

Final testing and truing up of six strip spline, when drawn through "V" notch of center gauge.

is flexed. A spline that is undersized, although it may fit perfectly into the assembly, will still be weaker than its neighbors, resulting in a rod having an erratic action.

Thus, if a poorly shaped spline requires very much reduction in dimensions to obtain a perfect triangle, and so a perfect fit, it had best be discarded and a new one planed from an extra strip saved from the same bamboo.

ABOUT HOLLOW CONSTRUCTION.

To fully "hollow-build" the larger rod sections, as is done by some of the fine custom rod makers, requires extremely careful work and the best of gluing techniques. There is comparatively little bamboo surface left for gluing in the hollow rod. Later you may wish to experiment with this interesting method, for it has great possibilities, particularly if you desire an ultra-light fly rod which still has great power and casting ability.

It is quite possible—and perhaps advisable—to build your first rod butt semi-hollow, by planing off the inside apexes of the splines, to leave a hole from ⅛ to $\frac{3}{16}$ inch after they are assembled. This work should be done after the final fitting and proper calibrations have been proven. The inner narrow portion of fibre planed away will not entail any appreciable loss of material, since this is the poorest part of the wood in the spline. Being less dense than the exterior, it will absorb more moisture from the glue, and would require longer curing after gluing, if the center were left solid.

This opening the length of a rod stick then permits better evaporation of moisture from the interior, and so assures more even curing of the glue during the later seasoning process. A smaller hollow may be used in fly rod middle sections, and for perhaps two thirds of the length in one-piece rods, but it will not be necessary or desirable in small tips.

With your set of rod sticks now completed, enter the final micrometer readings below each figure on your chart for later comparison and correction when the exterior enamel has been removed.

If, during the shaping, you have used the two-bar steel planing form, you will have planed the strips down almost to the required size with the bars set for the required taper, as described in Chapter 6. But then the micrometer would still be necessary to determine spline thickness as each strip neared completion, so there is little difference, after all, in the general technique of shaping the splines in either type of form.

Now comes the most important step in rod building, the gluing, every step of which is covered in Chapter 11.

CHAPTER 9

THE FIVE-STRIP ROD

This unique type of construction perhaps merits a fuller description than was undertaken in Chapter 2, since you may be undecided at the start whether to build this rod or the conventional six-strip.

Here we are departing from the traditional hexagonal design, which has served so well for three quarters of a century, so the merits of five-strip construction should be carefully examined. In addition to the bracing effect of the apex opposing the flat, arched surface and the off-center glue line, as mentioned in Chapter 2, there is the undeniable fact, easily noted in tests of a five- and six-strip rod of the same type, and containing the same weight of bamboo, that the five-strip has more power or backbone.

This difference is particularly notable in fly rods, and is evident in rods of the bait casting type as well. I have not had the opportunity to compare heavier sticks in the two types, but feel sure that in five-strip construction such rods will prove unusually resistant to "set," which so often deforms such rods subjected to terrific strain in fighting large fish.

Robert Crompton, a fine craftsman of the old school, with the enquiring, analytical mind of a trained engineer, is credited with this radical change in construction from the time-tried hexagonal rod, and he pioneered in this field a great many years ago, experimenting with various rod tapers which resulted in proved working charts. Through the years he carried on a wide correspondence with amateur rod builders throughout the country, passing on much valuable information, covering every phase of rod building technique. Thus, many sportsmen and amateur craftsmen became converts in the ranks of the five-strip enthusiasts, of which the writer is one.

Basic construction is the same as in six-strip work, and the only addition to the equipment needed will be the five-strip planing form, having four pairs of matching grooves, as earlier explained. Now we encounter something new in determining strip diameter to produce a rod of the desired type, perhaps one chosen from the foregoing charts.

In six-strip construction we have found it easy to divide an over-all calibration at any point to determine the strip diameter, measured from flat to apex. But this pentagon, with its odd number of strips, requires a more involved method of calculation. In this case the strip diameter is less than half the over-all rod diameter. For example, a rod "miking" .290 at a certain point would have a strip diameter there of only .130 inch, instead of half .290, or .145 inch, as in the six-strip.

The factor for determining strip size is 2.236. Thus, if we divide our .290 by this figure we get .130, or practically that, after dropping off the last decimal. It is not too laborious to thus chart the strip diameters for one rod you may wish to build, and you can easily make a universal conversion chart, suitable for all five-strip work.

For this, use a sheet of graph paper, divided into 3/8-inch squares, each of which is divided by four lines. For all rods up to an over-all diameter of .450 inch, which covers practically all rods for fresh water use, cut the sheet 17 11/16 inches wide, measured from the left base line, and 7 15/16 inches from top to bottom, measured up from the lower base line.

Use a fine-point pen to draw a straight, diagonal line from the lower left corner across to the upper right corner. Then, starting at the lower left corner, number the heavy lines denoting each square up the left margin, from zero to .010, .020, .030 and so on to the top. Between each of these figures number each fine line—

there are four—as follows: 2, 4, 6, 8, each of these representing an increase of .002 inch.

The squares and intermediate lines are similarly numbered across the bottom. These represent the over-all rod diameter, the figures on the vertical margin, the strip diameter. Thus, to find the proper strip calibration for a rod at any given point, follow up the sheet from the figure at bottom to where it intersects the diagonal line, then across on the lateral line to the left margin. The reading there will indicate the required strip diameter. The figures for an entire rod can then be entered on your chart, each just below the over-all rod calibrations.

In your first five-strip work you may well copy one of the charted rods in proved tapers, or copy the over-all dimensions of a favorite six-strip, and attain these in five-strip construction by use of the conversion table.

FIVE-STRIP PLANING TECHNIQUE.

You will start with your strips squared and straightened, just as for six-strip work, but they should be a bit wider on the face for a rod of given calibre, for reasons which will appear later. Already in the preparation you will be pleased to find that less time has been consumed in roughing out the strips, with four less pieces of bamboo to work up in a three-piece fly rod with extra tip, for example. And this will be only the first pleasant surprise.

Your five strip planing form, having grooves in matched pairs, will require that each strip be planed down in each groove of a pair. This is necessary because our strips now are not equilateral triangles of 60 degrees, as in the six-strip rod. The angles of each side are 72 and 54 degrees, necessary for perfect joining of the five strips to produce a perfect pentagon. This is a somewhat involved affair, with which we need not be concerned, however, since the grooves are accurately milled for the desired result, and we need only do accurate planing in them.

For some strange reason I do not pretend to understand the strips seem to work down better and easier and assume truer form under the plane than in six-strip work. In the final fitting, too, the five seem to lock in and fit up better, with a minimum of trouble, than would a corresponding set of six splines.

Again, as in six-strip planing, it is best to start with the rod butt. Also it is well to form the habit of starting to plane each strip with its enamel face to the wide side of the groove, which is on the left; this is merely to establish a custom, which you will soon unconsciously follow in all the work. Make a few short, overlapping strokes with the larger plane now, as in the six-strip work, then turn the strip over and into the matching groove, always with the enamel face to the wide side. Remove a bit of the upper corner, turn back to the first groove, and so on until it is down to triangular shape.

There is a bit of caution necessary here, as the strip takes shape. Since one planed face will be wider than the other, more cuts will be taken from the wide than from the narrow side, necessitating perhaps three strokes from the wide face and two from the narrow side each time the strip is turned. This is the reason for starting with strips a bit wider than for a six-strip rod. Also, of course, the completed outer faces of the five will be wider.

When the rough beveling is done and the little low-angle plane resorted to, it may be found best to take only one light cut on the narrow side of the strip and two on the wider face. When it becomes necessary to file down the nodes to prevent tearing, the strip will be found to lie level with the surface of the form by turning its inner apex down in the groove which fits its angle; this may be resorted to from time to time to test the accuracy of your planing. By sliding it down to be exactly flush with the metal surface, you can press it firmly in place and note if the upper, enamel face is perfectly level, without cant in relation to the surface of the form. Even more careful checking may be done by sliding the edge of a steel rule along the full length of the strip.

This checking, with the edge of the rule forming the third accurate side of the triangle, since it is flush with the face of the form, will indicate any point at which your plane has not been held true, for in that case one edge of the strip will be higher, indicating a "long side" of the planed face. This method of checking your work corresponds with the use of the "V" in the center gauge in checking the splines for a six-strip rod.

Of course you might proceed with all confidence, plane down all five strips for a rod section until they "mike" to exact diameters at all stations, then spiral them with string for a first fitting. This method will tell the true story, but until you have had some practice in five-strip work, it is risky, for a badly shaped spline cannot then be corrected without sacrificing some wood in the cutting. Then it will not match the others in size and so carry out the exact taper of the rod at that point, even though you do succeed in making the corners fit perfectly.

So, far better to make haste slowly as the splines come down close to the finished calibrations, and to use the micrometer often, as well as to check the angles, as described above. Always, with your first rod, and in succeeding ones, too, in which you hope to have included some *marvelous* new degree of taper to make this a "super" rod, you will want to hurry the work. But don't do it. Remember that the sticks must, at best, season a long time after gluing if you are to have a good rod. So why hurry now?

Always, as you work and fit and test the strips for angle, there will be a temptation to neglect some tiny imperfection, such as a slight gap near a node, which was torn by your plane being a bit dull—another sure sign that you are rushing the work. It will seem evident that the glue will fill that thin space along the corners, so it will not be noticeable. And the glue will fill it, quite adequately, but the imperfection will still be plainly visible. After the rod has been finished and varnished, there will be an unsightly dark line along that apex of the rod—damning evidence that you were careless and did rush the work.

It is better, always, when a spline gives too much trouble and just will not shape up right, to discard it and make a new one from a strip saved from this same cane. Proper attention must be given to cutting it the proper length, of course, to space the nodes with those in the other completed splines.

In this final fitting of the splines, as in six-strip work, a perfect job will be indicated if the moderate pressure of hand-wrapped cord brings them into perfect contact at every point. But, if this cannot be accomplished, try the set through the little pressure winding machine, which I sincerely hope you have purchased. It, next to an accurate steel planing form, is the most necessary item of your equipment to insure faultless gluing. Very often, then, a "dry run" of the stick through this marvelous little machine, which binds on the string at *constant* pressure—and at any desired pressure, while you merely turn the handle—will disclose that your fitting work has been good after all. Additional pressure—more than possible with your hands—has done the trick.

In the final fitting of the five-strip splines, the corners should be slightly sanded off, as in six-strip work, and the inner apexes flattened a bit, except, of course, in the areas where the ferrules will be mounted. The butts and larger sections may be hollowed as in the "six" to any degree you think wise, remembering always that if a hole in, say, a butt section is much larger than ⅛ inch in size, the loss of wood may require that the chartered diameters should be a bit more than for a rod which is solid to the core. If your desire is merely for a small hole in the center to hasten the seasoning of the sticks after gluing, and to remove that thin inner apex of coarse wood, this slight loss of material need not be considered as affecting the action of a rod calibrated as a solid stick.

There is now a final point which should be understood regarding your five-strip rod chart. If you will turn back to Chapter 4, and the charts, you will note that in a six-strip rod the inside diameter of the female ferrule is usually about that of the stick diameter at the end. Thus, in fitting the ferrule only the corners need be taken down until a perfect circle—the ferrule seat—is formed. None of the fibres of the flats will be taken off in any appreciable quantity.

But our five-strip has wider faces. If we use a ferrule equal to the diameter here, we will still have a reduced pentagon, with just the corners removed, and not a round, tightly fitting ferrule seat. Thus, it is better to use a ferrule 1/64 inch smaller than the stick diameter, for this reduction of .015 inch will just let the seat be accurately rounded and still not remove more wood than necessary. Large sticks, however, may require an allowance of .025 inch.

So now, having reached the same point in the new five-strip as we did in six-strip construction in the last chapter, we can consider the next step which, as in all the succeeding work, may be applied to either rod type.

CHAPTER 10

HEAT TEMPERING THE PREPARED SPLINES

At this point, with your set of rod sticks carefully fitted and ready for the gluing, perhaps you will be considering the possibilities of tempering them, and wonder how to go about it. Many manufacturers do treat their rods to add stiffness and power, and some of them advertise miraculous results achieved from their own "secret process." There may be highly scientific methods, perhaps, in the use of chemicals which will improve the bamboo structure, but such treatment is beyond our ken. We are here concerned only with what may be accomplished in the home workshop with simple equipment.

The usual method of bamboo tempering is done with heat, carefully controlled to gradually penetrate each section of the wood to its core, without an extreme temperature which would char the exterior. Under this treatment the naturally light tan, or lemon-colored structure, darkens to shades from dark tan to deep brown, depending upon the degree of heat and length of the treatment.

The color extends entirely through each section of bamboo and, of course, is permanent. But the color is immaterial, although the brown-toned rods are often of pleasing appearance, especially when finished with fittings and wrapping silk which are in harmony with the color of the wood.

There seems no question that brown-toned cane becomes much stiffer, due to the chemical changes which take place as a result of the baking. What actually happens in the process we do not exactly know, but it seems logical that the resins become more brittle and thus the fibres of the bamboo, being harder, are stiffer, more solidly bound together, and thus impart greater power to the finished rod.

Brittle wood, of course, will fracture more easily, so the problem in heat treating is to find the critical point, in time and temperature, to achieve improved stiffness without sacrificing materially the toughness and resistance of the wood to severe

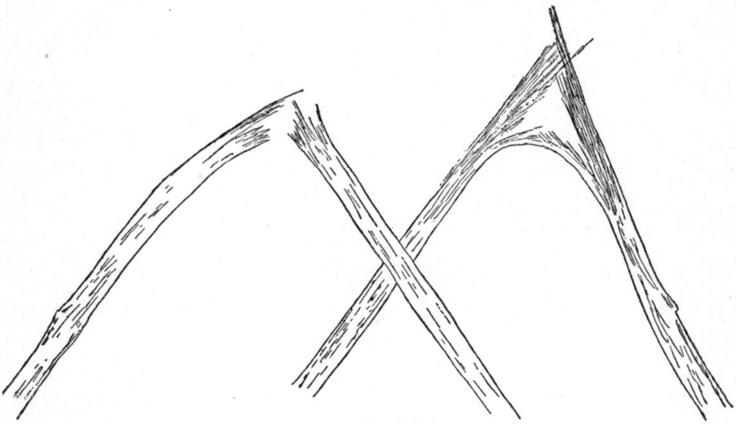

RIGHT: *Good bamboo fibres will break only with a long, splintering brushlike effect.*
LEFT: *Coarse fibres, or those too severely heat treated will break almost squarely.*

strain. A section of bamboo, heated to dark brown can, when cool, be broken square off in the hands. A similar stick in its natural state will "let go" gradually, under the same pressure, separating in long, brushlike fibres, the enamel or outside portion of which can almost be tied in a knot before the separation is complete.

Mr. N. Uslan, manufacturer of the beautiful five-strip rods already referred to, outlines his ideas as to heat treatment much as do the above two paragraphs. He believes that if it were always possible to use canes of the very finest quality, heat treatment would not be necessary; we might put it in other words: that we cannot improve upon nature's best work. Thus, we can readily understand that commercial rod manufacturers are confronted with the problem of sorting and inspecting large shipments of bamboo stock, perhaps reserving some canes for use in their natural state; while others, still good but not of the highest quality, can be improved by correct heat treatment.

There is no question in my mind that many rods in the low-priced field are made of severely treated bamboo which was of mediocre quality in its natural state; and that thus a useful rod, certainly "worth the money," can be built from inferior canes. In the finer rods, from smaller shops, where careful testing and sorting of the canes and resulting strips is possible, we may feel sure that every angle of the tempering process is carefully considered.

Mr. Lew Stoner, guiding genius in producing the Winston Rod, in San Francisco, thinks that heat tempering perhaps increases a rod's stiffness and power 25 per cent, but still he frowns upon the practice of excessive heat tempering, with the possibility of brittle wood and resulting breakage in the hands of a careless angler, who then, of course, would blame the maker. He strives, instead, to incorporate power and fine action, coupled with strength, in a careful study of rod tapers and in faultless workmanship.

Perhaps all our finest custom rod makers would depart farther from many accepted practices and incorporate methods which would provide "the perfect rod," if only they could be assured that *every one* of the beautiful creations over which they had labored would have proper care. Certainly, the careless angler who has rod trouble and always blames the maker must be anathema to the manufacturer.

Other fine craftsmen throughout the country probably disagree violently as to heat treatment of bamboo; and, if the major premise of its value is accepted, their ideas will differ widely as to methods and the proper degree of heat to use. The subject is inexhaustible when all the relating factors are considered. Perhaps science will some day give us a "magic" chemical which can be simply applied and will produce marvelous results.

Robert Crompton, after thorough and objective research in every phase of rod building, was opposed to heat tempering in any degree. His approach to the problem was that of the scientist, intent on producing comparatively few of the finest possible rods, regardless of the time consumed in studying, sorting and testing the bamboo strips.

Heat treatment or tempering, requiring a comparatively high degree of heat, should not be confused with seasoning of bamboo stock in moderate controlled temperatures, as most manufacturers probably do to expel most of the moisture before the stock goes into rods. In this process—and you may find it necessary—your half canes or split strips are simply stored in a warm, dry place for a long period. To speed up the curing process you can easily construct an adequate and inexpensive shop heater, such as the type described in Chapter 12.

But now we are concerned with actual heat tempering, which surely will improve poor and mediocre bamboo stock and may improve even good material when the heat is carefully controlled and the proccess carried only to the point where the bamboo changes in color to a mild tan or very light brown shade.

The tempering should be applied only to the beveled and finished splines. While it might seem desirable to treat the squared sticks before shaping, the treatment renders the wood so hard and tough that later planing is difficult. So, if you wish to temper, use a section of galvanized iron pipe about 1½ inches in diameter and 6 inches longer than your rod stick.

Shape a soft-wood plug to fit snugly into each end, and drill each with a ⅛-inch hole for the escape of vapor. The pipe may either rest in a simple frame made of strap iron, or be suspended from the ceiling by two wires looped to hold it at a convenient height while you apply the flame of a blow torch along its length.

Next, procure a few dozen cubes of *ammonia carbonate* from

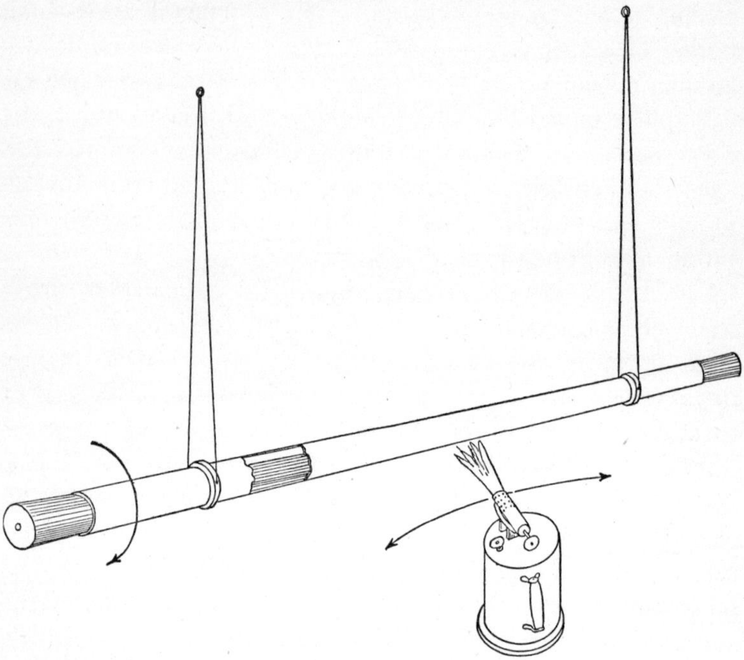

Galvanized iron pipe used for tempering rod splines. Pipe is slowly revolved while blowtorch flame is played along its length.

your druggist. We are going a bit farther than plain heat treatment, you see, and employing a little "secret" process, which for me, at least, has given better results than with heat alone.

The pipe will hold several sets of rod sticks, enough for two or three rods, if required, and this many may as well be treated as one set alone. But have each set of splines marked carefully, then each individual spline marked, so that none can possibly be placed in the wrong rod or position for gluing. You will no doubt evolve your own method of numbering the splines. But use a lead pencil for this work. The marks will endure under the treatment, while ink may run and blot.

Load the pipe with the bundled sticks and insert a cube of the ammonia about every 3 or 4 inches along their length, by placing it on top of the strips, carefully shoving them a bit farther in, adding another cube, and so on. About a dozen will be sufficient for perhaps six or eight rod sections, and a few less will suffice for a single rod. Care should be exercised that the cubes remain in position as the sticks are finally shoved full length into the pipe. Your wire loops, too, should be evenly adjusted to prevent tilting and possible rolling of the cubes when you turn the tube.

Now, with the pipe suspended and the end plugs in place, start your blow torch, obtain a strong, blue flame, and slowly apply it the full length of the pipe, which should be slightly turned every few seconds with a gloved hand grasping one of the plugs. Continue this heating without once stopping the movement of the torch. The idea is to heat the entire pipe slowly and equally. The constant revolving of it in the loops will, of course, aid in this heat distribution and, equally important, keep the rod sticks slowly tumbling and changing position, thus preventing them from receiving direct heat from contact with the metal. This occurrence could transmit more heat to one rod spline than to another and might, should you stop the movement of the torch, slightly char the delicate, beveled corners. Actually we are heating that bamboo stock with super-heated air, you see. There is none of the haphazard "flame-tempering" you may have read about in the advertisements.

You will be wondering just what is occurring in your oven, after five minutes of steady treatment; but do not bother to look—just keep the flame moving and the pipe slowly turning. Soon now, a thin coil of damp vapor will curl from the vent holes in the plugs, indicating that the wood, which seemed so dry and bonelike, still possessed a little moisture, which might just as well be eliminated. This may not indicate that the wood was not well cured, either. It can be due to absorption of moisture from the air, in climates where the humidity is high.

After ten minutes the little steam jets will puff more strongly, then slightly lessen, and you will detect the odor of the ammonia.

This will indicate that practically all the moisture has left the wood, that the heat is melting the ammonia cubes and releasing the strong fumes which now are penetrating the open pores of the bamboo. So our "secret" chemical treatment is progressing nicely. The ammonia vapor now will curl from the vents for five minutes or more, so continue the heating until it lessens in volume, and you find the odor—after a very cautious sniff at the vent hole—is being replaced by a faint hint of "hot wood."

Now, with gloves on, remove one plug and draw out a few of the splines for inspection. If no traces of the ammonia cubes remain, and if the wood has turned an even shade of tan, straw or light brown, the tempering is complete. If the color is not changed materially and you are anxious to have one of these often desired shades, the sticks may be returned for a few minutes and the heat turned on again. But, beware of continuing too long and ending up with bamboo of a brown tone, for then it will have reached the dangerous, brittle stage.

After all, the color is immaterial, except that a moderate change, say from a natural lemon shade to a rich tan or straw, will indicate sufficient heat to attain the desired result in improving the fibres of the bamboo. After the torch is put aside it is well to leave the sticks in the pipe for an hour, to cool gradually along with the metal. This will permit evaporation of any moisture which may still be present.

If your stock was not fairly well cured before being worked up, you will not only find steam coming from the vents for a longer period, but an actual trickle of water when a plug is removed and the pipe tilted slightly. Thus, it might be well, if you are uncertain of its condition before reaching you, and you cannot wait for a period of slow curing, to insert a few odd sticks in the oven, heat it well, and observe its moisture content. Such practice also will serve as a "trial run," and insure more confidence in the final treatment of the splines over which you have labored so long and diligently.

Various other tests in the oven are interesting and instructive, too, such as heating two squared strips of similar dimension, one to the "right" degree, the other left in the pipe without turning it, until it is a deep brown. Then break each strip between the hands. The first should fracture progressively as strong bending pressure is applied with the inner, pith surface uppermost. After it has separated, the stronger outer fibres will let go reluctantly, a few at a time, and "brush out" like the bristles in a paint brush. The brown stick will probably break clean off, with the same pressure, for its fibres have simply had all the life "cooked" out of them.

So, by observing the cautions above described, and following the general technique which I, at least, have found safe and satisfactory, you will have a set of nicely tempered sticks, in which the ammonia has in some magical manner replaced the lost moisture, and thus hardened and toughened the bamboo fibres.

The splines, at last, should be laid apex up, in sets, and the planed surfaces briskly brushed with a set of stiff bristles, such as in a worn-down paint brush. This will remove any dust or ammonia residue and leave the faces clean for the gluing job.

CHAPTER 11

THE ALL-IMPORTANT GLUING

For many years, before a scientific gluing—or pressure winding machine—was available, the home workman could use only crude, hand methods, which were unreliable and nerve racking. Before the advent of the excellent cold-water glues we have today, most rod gluing was done with hot "animal glue," which, for best results, had to be kept at a certain temperature. The work-room and the rod splines also needed to be warm.

Such glues, even when properly applied, were always subject to heat and moisture in a finished rod, and were responsible for much of the "set" so often developed in the older rods. The humidity and changing temperatures during one day's fishing would sometimes change the rod's action and develop a sluggish response. This phenomenon, while it need worry us little in the use of modern glues, still demonstrates that varnish is not absolutely impervious to moisture, although it does offer considerable protection.

It is still possible to assemble a set of rod splines, join them with modern glue brushed on the surfaces, and apply the wrapping string by hand, as described in Chapter 8. A reverse spiral wind, of course, is then necessary to remove torque. But there are many difficulties in this crude method. The wrapping pressure will be inadequate and variable. Corrections for torque and kinks in the stick will be difficult to make. I have glued up many rods in this manner, but it is wholly unsatisfactory, and the purchase of the little pressure winding machine before mentioned is highly advisable.

Note the illustrations on page 77 showing a similar machine

clamped to the edge of the workbench. A narrow shelf, longer than the rod section to be glued, is fastened flush with the entering "U" of the machine, where the stick starts; and a similar shelf, or a half-round trough, made of a half-section of two-inch galvanized pipe, leads the wrapped stick from the machine in a straight line. Thus, it is supported throughout and goes through without being warped, in spite of the great pressure exerted upon it.

The detailed drawings indicate the action of the endless belt in driving the stick through, show how the belt is looped twice around the stick, with the wrapping string caught under the first turn, and how the reverse wind is made. It is wise to practice a bit with the machine and an old rod section in order to become familiar with the various simple operations before you attempt the actual gluing of your new rod.

ON GLUES.

Happily, with the day of hot glues past, we now have the fine caseins, all in powder form and easily mixed with cold water. These are water-resistant to a high degree, if not absolutely waterproof, and will endure a moderate degree of heat without harm, a valuable feature, which will be discussed later. I have had excellent results with Casco casein, which may usually be found in any hardware store. There are probably others, equally good, with which I am not familiar.

Some of the lately developed synthetic resin glues are also adapted to rod work. They hold tenaciously, are waterproof and those I have tried have proved highly heat resistant. Both Cascamite and Weldwood are glues of this type and may be found in cans of various sizes in most hardware and paint stores. These work very well for bamboo gluing and are also excellent for making rod grips of cork rings. The Herter Company, referred to earlier, is offering a new synthetic resin glue also, which comes in liquid form, with a hardening agent to be added. I have recently had excellent results with this adhesive.

Considerable experimenting may be necessary to secure always perfect results in gluing, and best results will usually be obtained by using one type until you have learned all its individual characteristics. Bamboo varies in density, the very hardest sometimes seeming not to provide a surface for a good glue "bite." The various glues will each work best with weights on your wrapping machine which exert a pressure in keeping with the glue consistency. In other words, a thick glue will require more pressure to bring the splines into close cohesion and force out the surplus glue, leaving the thin film which will insure adequate bonding and still not show a glue line at the rod corners.

In nearly all cases, with the powdered resins and caseins, follow the directions on the can as to mixing, and weigh both powder and water in the correct proportions. Resin glue requires a smaller proportion of water than the casein type, and may be used immediately after mixing.

In mixing casein glue, use a clean cup or small bowl, weigh into it the required amount of water, two ounces perhaps being enough for the gluing of one rod. Then add the correct amount of powder and stir it in until the mixture becomes thick and pasty. It will take about fifteen minutes for the powder to absorb the water completely, after which a final stirring will reduce the small lumps, leaving the mixture about the consistency of thick cream—and "plenty sticky." Then it is ready to use.

This is probably the best glue for the amateur builder to start with, and it will produce consistently good results when used exactly according to directions. With all its advantages, however, the rod sticks will require longer curing than when a resin glue is used, since the casein mixture contains more moisture which must be evaporated. New and better glues of marvelous adhesive qualities are continually appearing on the market, and you may well investigate some of these as time goes on, for comparison with those now available.

A wise procedure, when using casein glue, is first to treat each planed surface of the splines with a solution made of one part sal soda to ten parts of water, by weight. Dissolve the soda and just paint on the solution with a small brush, when the splines are laid out for gluing, as described below. When they are dry, brush off any adhering soda crystals and apply the glue. This treatment will neutralize any acidity present in the bamboo and thus prepare it for the glue, which is alkaline.

Now start with your largest rod section, the splines properly placed as numbered, roll them together, butt ends flush, and bind at each end and the middle with a strip of Scotch tape, or other waterproof adhesive. With a razor blade, slit each binding along one apex and lay the strips on the bench, enamel side down. The tape will hold them in a neat row to receive the glue on the upturned faces.

A stiff brush should be used for applying the glue and should be kept well filled, literally dripping the sticky stuff while you make long strokes, bearing down to work along the lower edges of the splines. Be absolutely *sure* that every part of the faces is well covered, and do not worry about the surplus on the bench top. It will wash off later. A brush about one-half inch wide is best, and I have found the stiff little brush sold for cleaning typewriter keys to be ideal.

Roll the strips together—the glue will hold them—take off the strips of tape, and the stick is ready for the winding machine.

USING THE PRESSURE WINDING MACHINE.

It will be noted that the slide on the machine, which carries the pulley and crank, weighs perhaps a pound; this alone will exert some pressure on the stick when the belt is passed twice around it, but this weight is not sufficient. It is necessary to use enough weight, for the size of stick being glued, to squeeze most of the surplus glue from the seams, leaving only a thin film.

Thus, a fly rod tip will require about five pounds of weight for the first run, a fly rod middle section perhaps eight pounds, and a butt section ten or twelve pounds. Engineers compute that ten pounds of weight on the slide will exert a pressure of two hundred pounds on the glue faces. Considering this, do not overdo the weight and risk having a "starved" joint from which all the film has been squeezed.

The endless belt, provided with the machine, is of closely braided material, and should last for some time. Others may be purchased, or they can be made from a length of one-eighth inch cotton chalk line, if you can manage a long, smooth splice in the three strands.

Wet the belt, before starting the stick through, to render it more pliable and to provide a better grip on the bamboo. For wrapping, use ordinary grocer's No. 8 cotton string, either placed in a bowl or wound on a spool which has been mounted on a dowel let into the bench or the machine base. In any case, it should render through the tension plates of the machine freely, but without slack.

The string supply should be at right angles to the machine base, so that one complete turn around and between the plates will actually make one and one half turns to where the string leads through the guide to the rod stick. It is well to test the tension by pulling on the string while the tension screw is slowly closed down. With too much pressure the string will break during the winding.

Most of my own gluing has been done with a machine built after Robert Crompton's original pattern, and the Herter machine is also somewhat similar, although it varies in some respects. The illustration is made from the latter, but with the two central pins which guide the belt replaced with a single unit having two "U"-shaped notches to guide the stick. The space between these accommodates the central turn of the belt, and governs the width of the spiral windings to about three-eighths inch. The use of the machine, however, is identical with one having the two round, guiding pins.

Pressure weights, to be hung from the hook on the slide, may be chunks of odd metal of the required weight; or even a small

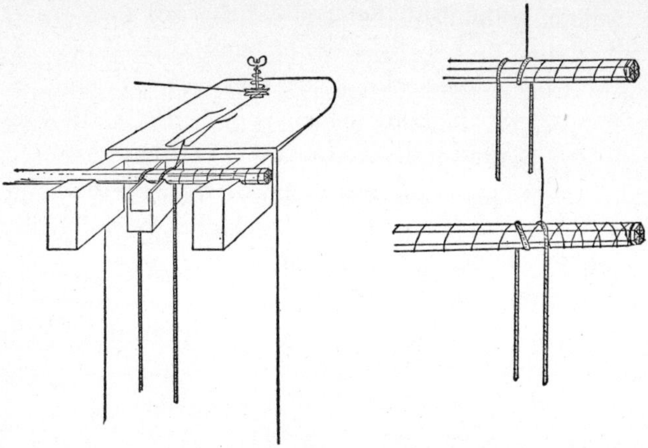

Starting glued rod section through the pressure wrapping machine. Cord passes between tension plates, adjusted by wing nut, over guide and under belt, which spirals it evenly.
Right: Above, belt applied for first wrapping. Below, Reversed for second wrapping.

pail of sand may be employed, and the contents added to or reduced for rod sections of varying diameters. I have found that discarded household irons of the old type, weighing from three to six pounds, will serve nicely and permit a quick selection of the necessary weight. It is wise to have all required equipment ready at hand before starting the glue job, or the adhesive on the treated sticks may start to set before full wrapping is accomplished. Therefore, have a pan of hot water and rags ready before starting the first stick through the machine.

Grasp the butt end of the stick in the left hand—very soon after spreading the glue—and place it in the central notch, as shown in the illustration above, with just the end projecting, but not reaching the right-hand guiding "U." Take a double turn of the belt around the end, with the central turn between the two fingers, or pins, this turn being made from front to back of the stick; that is, the last turn of the belt will lead down to the pulley from the back of the stick.

Pull the winding string snug under this last turn, and tie at the butt of the stick with a double half-hitch. Then, as you steadily turn the crank clockwise, the stick will be driven through, tightly spiraled by the cord, which is guided accurately by the belt at the right-hand pin. Even pressure will have been exerted throughout the operation, and you will note the glue literally oozing out from each apex of the stick. But now it will have torque, or twist, due to the spiraling under pressure.

Wipe off most of the adhering glue with a rag dipped in the hot water and squeezed out, cut the string at the end, and tie off with another double half-hitch. Following this operation, inspect the stick for glued faces not tightly closed. Such an occurrence is unlikely if you have observed the three precautions mentioned earlier: Proper glue mixture; sufficient weight on the slide and strong tension on the winding string. There is considerable latitude in these matters, in relation of one to another. Thus, you might use a very thick glue mix, which would require a heavier weight on the slide to press out the surplus and bring the wood faces into close contact. But this is not advisable, especially if you are employing one of the caseins.

If, for any reason, the stick is not properly joined, strip off the string, separate the splines and at once wash the glue from the surfaces. While they are drying check carefully each of the above important points, and reglue this offending first effort.

The second run through the machine will remove that twist, which has been worrying you, and add additional pressure to the splines with the second wrapping, which is reversed to the first. This time wrap the belt just opposite, with the last turn leading down to the pulley from the outside of the butt of the stick. Pass the string under, as before, and tie off.

Now considerably more weight will be required, to counter-twist the stick and remove the torque. Double the first weight will usually do the trick, but there is no exact rule applicable, and sometimes, particularly on butt sections, being stiffer and of course more resistant to the twisting effect, only fifty per cent more weight will be required. A bit of experiment will soon teach you the approximate weights for your own work.

It will be found wise to wash the belt in the hot water each time a glued stick is wiped, to keep it from slipping when impregnated with glue. Finally, it should be well scrubbed with soap and water to preserve it, smooth and flexible, for the next job.

If you have been careful—and lucky too—in this first gluing operation, the sticks should come out straight and without torque, or twist. Butt and heavy middle sections should give little trouble. With the slender tips you may not be so fortunate.

So sight down each stick at once, while the glue is still soft, and correct any slight bend by careful counter-bending with the hands. Too strong effort will make the matter worse, by imparting a new kink, so work gently. When each stick appears perfectly straight, after careful sighting along its length toward a good light, lay it on its flat, and examine for torque. Countertwist the stick, or the area affected, back to a true plane, and leave the section for a few moments to see if there is any tendency for it still to draw back as it was before.

A period of about twenty minutes will permit this work, before the glue begins to set and prevent easy manipulation, so work briskly and carefully. Here, if torque has proved troublesome, is the time to learn *why*, for future reference. A twist in the direction of the first run will indicate that too little weight was used on the second run to counteract it. An opposite twist will be due to excessive weight on the second passage of the stick through the machine.

It is wise, after the sticks are all glued, to make notes of the correct weights used for each rod section and to have these for ready reference for later work.

Longer rod sections, such as for two-piece fly rods and one-piece bait rods, are a bit more difficult to glue. Bends and torque, however, may be corrected in the manner described above. With these longer sections, "drag" may be imparted to the slender small end due to the friction of the larger end as it revolves in the receiving trough. This should be kept bright and smooth, with all adhering glue washed off. In a pinch, if unusual drag persists, have someone help you by holding clear and revolving this heavy end of the stick, at least during the last part of the first run.

Another method of straightening glued sticks is to have a smooth, *perfectly flat* surface, such as a strip of heavy linoleum, laid on the bench top, and to slap each section smartly upon this. But the trick is to have *all* the stick strike at exactly the same instant, or there will be more bends. This is not so easy to do. Some professional rod makers use a long block of smooth stone for such straightening; and they have plenty of practice, which we lack.

It is far better, I think, to make every effort to have the sticks come from the machine perfectly true than to rely upon any special treatment to correct them afterward. The stick which can be sighted down, found true, and immediately run through the wet rag to take off most of the exterior glue, then laid away without further attention, is the one which will make the best rod.

That final "wet wipe," also, will add even more pressure to the wrapping string, for the dampness will shrink it even tighter. This is the principal reason why ordinary cotton is better than linen or any of the treated strings or heavy threads, which otherwise might work very well, and perhaps be stronger.

CHAPTER 12

SEASONING AND FINISHING THE GLUED STICKS

You will wish mightily—every day—to have that first rod finished, fully fitted and ready for use. But now patience becomes a real virtue, for the sticks will require slow seasoning for the best results. However, you need not worry about waiting for six months or a year, as the best professional rod makers were said to do in the past when artificial heat curing was not in use.

It is still possible that natural curing in normal air temperature might be best, but with modern glues there is no reason why we cannot take advantage of moderate artificial heat, if the increase in temperature is gradual. Thus, it is good practice to first hang the rod sticks vertically, from nails, along the shop wall for a week or so. Transfer to somewhat increased heat is then permissible. Hanging beside a hot water heater of the automatic type will do nicely. Adjacent to a basement furnace is good—if a thermometer test indicates heat no greater than 80 or 85 degrees.

The sections should always either be hung vertically or laid upon a solid, flat surface. The latter method, however, does not permit an even distribution of air, as when the sticks are hanging free. After another week remove the wrapping cord, lay the sticks flat upon your workbench, and remove the adhering glue along with the thin layer of enamel. A sandpaper block, No. 0 grit, will serve *if* you stroke lengthwise and are careful to keep the block level with each flat to prevent rounding the corners. A small, hooked-blade carpenter's scraper is excellent for this work, also. It should be kept very sharp and used with easy, short strokes, which will take off most of the exterior, hard enamel barely down to the outer power fibres.

This enamel could have been removed while the splines were being planed, to aid in obtaining more exact calibrations, as explained in an earlier chapter, and you may prefer this procedure. In this event the cleaning-up now will require only a light scraping to remove the film of dried glue. Finally rub each flat of the stick—always lengthwise—with No. 4-0 garnet paper or an equally fine abrasive, and apply it lightly, after which the corners may be slightly dressed off—just enough to remove the "hair edge."

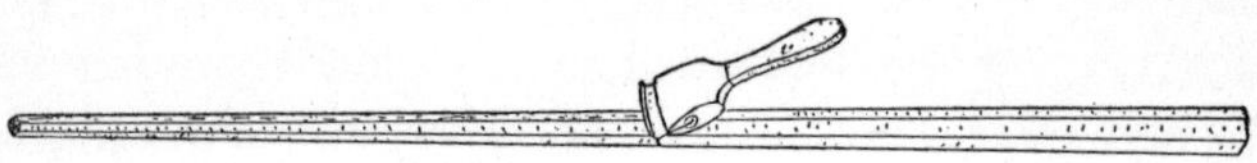

Method of using hooked-blade scraper to remove exterior enamel from glued rod sticks.

The last operation is to use a pad of No. 00 steel wool to polish the stick. This will produce a velvet smooth surface, without removing any of the fibres, and permit close inspection of both the outer grain and your fitting and gluing job. If you have been extremely careful throughout—and exceptional luck has been with you—not a single glue line should be evident at any point. The sticks will appear to be solid wood, without any sign of joining.

But these are your first rod sticks. You have undertaken a difficult job in both fitting and gluing, and absolute perfection is not to be expected. So never mind if an infinitesimally fine brown seam at some spot indicates a faint glue line. Just console yourself by examining carefully any commercial rod of the medium-price class, a "machine-production job," and compare with your own handiwork. You will feel better. Test calibrations with the micrometer will almost surely prove that your splines are truer in cross section and more accurately tapered.

Already you will have sighted down each stick for straightness, and perhaps have discovered a slight bend or two. This may be due to the wrapping string having prevented a true appraisal after the gluing, and such bends will most likely appear in tip sections. Again, if the bamboo was not well seasoned before gluing, slight warping may take place in any stick, due to the effect of moisture in the wood during the post-gluing period.

Also, stresses may have been set up during the drying, if the inner apexes of the splines contained an area of coarse fibre, with its interlying pith. This being more porous than the hard, outer fibre area, will take in more water from the glue, require more time to dry, and so may possibly produce a tendency for the stick to warp. Additional hanging in moderate temperature may cure this. The method of partly hollowing the larger rod sections, as described, with bare removal of the sharp apexes of the tip splines, should largely obviate this trouble.

So do not worry about a warped stick now, for it can be corrected later, after full curing is completed. With two weeks, say, having elapsed since the gluing, the sticks will need at least a month more before finishing. Again, constant temperature of from 75 to 85 degrees is probably best, for at least another two or three weeks.

This drying period will depend largely on the humidity in the air. In localities where average humidity—moisture content of the air—is around fifty or sixty per cent, a longer curing period will be required than in a desert atmosphere, for even heat does not entirely dispel moisture. So again, it would be best to hang those sections back in a warm, dry place for several months, and proceed with the work on your next rod. If you just cannot wait, there is still a solution, the installation of a simple and inexpensive drying oven.

Professional manufacturers, I believe, use large ovens for curing their rod sticks, with the heat thermostatically controlled. You may some day wish to install such an oven on a smaller scale, but it is not necessary for home rod work. One built horizontally will take up considerable shop space, and perhaps prove difficult to heat evenly. My own, of the simplest possible design, serves admirably and costs little. If you wish to duplicate it:

Have a tinsmith build you a cylinder of medium-weight galvanized iron, about 7 feet long and 8 inches or more in diameter, with a long panel, about 5 feet long and 8 inches wide cut out of one side, for a door of the same material. This is hinged at one side and fastens with a hook at the other. The heater can be mounted vertically along the shop wall, or in a corner, with strap iron brackets.

An electric heating coil would probably work well, if designed and installed to give the proper degree of controlled heat, but considerable expense and work would be entailed. My own is mounted 8 inches above the cement floor and over

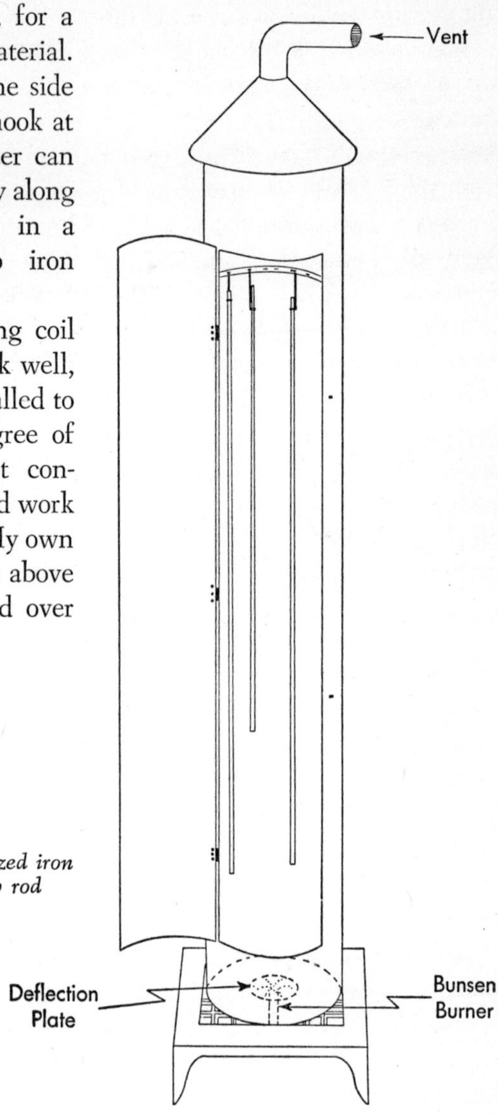

Simple heater of galvanized iron used for curing glued up rod sticks.

a little Bunsen gas burner, fastened to an iron stand—the framework of a castoff "one-hole" gas plate. A thin iron baffle plate above the burner diffuses the heat and prevents danger of scorching the lower ends of the suspended sticks.

Ventilation is assured by fitting a large tin funnel over the top of the cylinder and leading a pipe—or even a piece of garden hose—from the small end of the funnel through wall or roof to the outer air. The rod sections are hung by looped string from small hooks set in a strip of wood 5 inches wide which is fitted over the top of the cylinder under the funnel. This strip, while not blocking the draft entirely, still backs up the heat sufficiently to maintain a fairly even temperature from top to bottom. The air spaces at either side still permit fumes and escaping moisture to pass off.

A blue flame less than an inch high, from the burner, will provide sufficient heat, which is best determined by suspending a thermometer from a wire inside and taking readings at top, center and bottom of the heater. The average should be about 80 to 85 degrees for glued sticks which have had two weeks or more of seasoning at milder temperatures. It is best to keep the thermometer hanging inside, to permit daily readings and correct any possible change due, perhaps, to varying gas pressure.

Curing at the above temperature for a week to ten days often seems sufficient for tips and smaller rod sections, in my own experience. Thorough tests by severe flexing and final, more severe work put upon the finished rod has seemed to indicate that the splines had been sufficiently cured. Yet—more seasoning without the heater might have been better. If you live in a warm climate, having low relative humidity, natural air-curing would seem best.

Living at the seashore, as I do, the air is always damp, and with humidity of from 40 to 60 per cent, and sometimes more, it is unlikely that natural curing, without some artificial heat, would ever be possible. Thus, we might expect a glued stick to lose a little moisture during a sunny day; but it might actually gain by absorption from the air during a chill, foggy night.

Another method of slow curing, advised by Robert Crompton, might be worth trying. After a week or two he "cleans up" the sticks and lays them in a box of dry sand in his attic, where the sand, probably warmed by the sun on the roof, helps to draw the moisture from them. He sometimes left these glued sticks there for months.

There is another reason for thorough seasoning, too, aside from the major necessity of obtaining the best possible results from the glue job. Ferrules must be fitted exactly to the wood if they are to remain tight through years of use. Thus, the wood should be so well seasoned that it cannot later shrink even a bare one-thousandth of an inch. Such an occurrence and a slipping, rattling ferrule will not be fatal to that new rod, and it can easily be "fixed," as later explained, but it can be annoying.

The one certain method, so far as I know, of determining when the glued sticks are properly seasoned, is to weigh them. A little photographer's scales, weighing accurately to a couple of grains, will serve nicely. Or, your corner druggist may be willing to oblige. Thus, you weigh a butt section after the wrapping string has been removed, and find it weighs, say, 2½ ounces. After two weeks more, on its nail back of the water heater, it will probably lose another 30 or 40 grains. This represents quite a little moisture, you will agree.

After each weighing, mark on the butt the date and weight, and again return the stick to its place, perhaps this time to the shop heater, at a constant temperature of something like 85 degrees. A few days there, and the stick should lose perhaps 10 grains more. It is still curing, but more slowly, and another lapse of a few days may disclose no further loss of weight. Then, you may be sure it no longer contains any appreciable amount of moisture.

If, in the foregoing pages, I have seemed to stress as absolutely essential all the careful steps in rod seasoning, and you are confused and unhappy—having rushed that first rod to completion—

just have faith. While it is bad practice and would outrage the professional craftsman, I have used a rod fully completed two weeks after gluing. This was by way of experiment, and that rod did not come apart, either. In fact, it performed quite creditably on large steelhead. But, by every rule of science and common sense, the practice is unwise.

Now, with your rod sticks finally cured, you will examine them apprehensively for the least sign of warping, or slight bends resulting from the gluing operation. Now you can straighten any offending portion of a stick without harm—granting that you use due care. The application of mild heat will still soften slightly the fibres of the wood, just as it did when you straightened the split strips. But it will not harm the glue, as would have been the case had you used heat before the glue was well cured and while it still contained moisture.

So we have still another reason for thorough curing, assuming that sometimes we will have a stick that is not quite true. It is best to use heat from a closed-top stove, such as a small circulating-type heater, or use a metal baffle plate over a gas burner, turned low. Do not wear gloves, for your bare fingers will prove the best "thermometer" to govern the temperature. Pass the rod section back and forth over the stove, turning it constantly, and do not subject the wood to heat which is uncomfortable to the fingers. Quickly sight down it and carefully counter-bend against the warped portion. After a moment to cool, examine it again, and if it is straight, the job will be permanent.

But *always* use heat cautiously and proceed slowly, applying just the minimum amount which will permit the rod section to be slightly bent; that is, actually bent rather than given a natural arc. More heat will be required to straighten a large section than a tip, of course, but seldom should the heavier sticks need attention. The small part of tips will need only very slight *warming*.

If, in final examination of this first set of rod sticks, you find them straight and true, it should be cause for great rejoicing and a fast sprint into the house to proudly show your wife this final result of all the midnight oil you have burned. Your joy now, that satisfied feeling of great accomplishment, can only be overshadowed by an even greater event—the actual use of your new—and own—bamboo fishing rod.

At this point, while the sticks are bare of fittings of any kind, it is interesting to weigh the set, minus the extra tip, of course. This will give you the true weight of the rod for later consideration with its action when it is completed. As mentioned earlier in this work, over-all or completed weight of a rod indicates little or nothing, since reel seat, hand grasp and ferrules combine to form a widely variable factor. For a true appraisal of the actual "working wood" in the rod, weigh the sticks after each is trimmed to its exact length for the ferrules, as described in the next chapter.

It will be surprising to find that a nine-foot fly rod, for example, which completed will be in the 5½-ounce class, will actually contain 3½ ounces of bamboo or a bit less. The sticks for a husky little 7½-footer will weigh under 2½ ounces, and the tip alone for a three-piece 8-foot rod will weigh barely over 100 *grains*. These random examples, I think, will help us appreciate the marvelous possibilities of strength and power contained in the bamboo fibres.

Again, the weight of a small tip, in particular, will indicate how carefully the strips must be tapered to produce the proper action; how a few thousandths in diameter of the finished stick can affect its utility, considering the small total weight of wood to work with.

It will be interesting to keep shop notes of the weight of various sticks for later comparison with those in the same general class, but in which different tapers are employed. Bamboo density may make a slight difference in weight of sticks built to exactly the same specifications, but nevertheless the comparisons of total stick weight to final rod performance will prove valuable.

CHAPTER 13

FERRULES AND FITTING THEM

Few anglers and perhaps not many amateur rod builders realize the importance of good ferrules, properly fitted. Next to the gluing, tightly fitted and properly aligned ferrules—of the best possible quality—are most essential for the best rod performance. A ferrule loose in the slide is an abomination, as every angler knows, and one which loosens on its seat on the wood is not only annoying, but it can prove the means of ruining a rod section.

In considering the importance of a ferrule, remember that, particularly in a fly rod, the implement flexes and strains countless times each fishing day. At each ferrule, of unyielding metal, the torsion is suddenly interrupted. Both a twisting and bending action centers on the ferrule, and particularly where the flexing wood enters the metal this action is very severe. This is the principal reason why high grade ferrules are tapered to a thin edge where the open end fits the wood, and why this "featheredge" is either split or serrated, to partially relieve the strain at that point.

This taper at the butt end of each ferrule also permits final winding in finishing the rod, from the wood up over the metal, without a noticeable break or "step." The depth of the cut on the ferrule lets the wrapping thread end at the little shoulder, flush with its body. All these points, while minor, perhaps, will contribute to a more neatly finished and durable rod.

So, in ferrules we should consider both design and material and be willing to pay a substantial price for the best obtainable. Most of the best ferrules have been made of so-called German silver, an alloy containing up to 18 per cent nickel. This material machines nicely to a beautiful friction fit in the slide, is strong, and gives little trouble with the two sections "seizing." I have had splendid results with these ferrules sold by the Orvis Rod Company of Manchester, Vermont.

Fig. 1 illustrates the Orvis Ferrule in size 17/64, about right

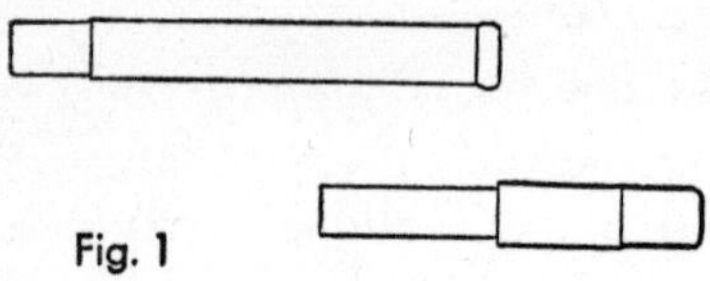

Fig. 1

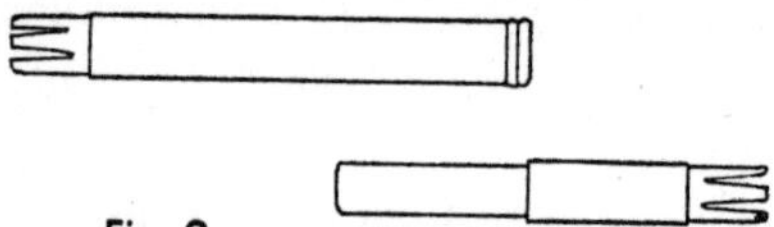

Fig. 2

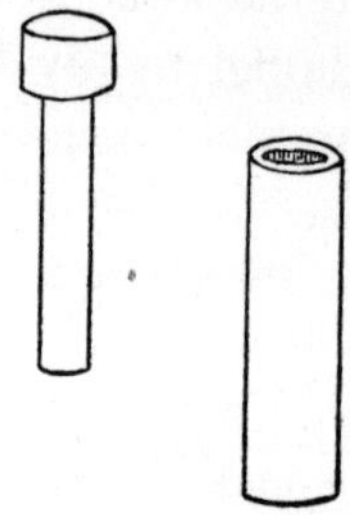

Fig. 6

Fig. 3

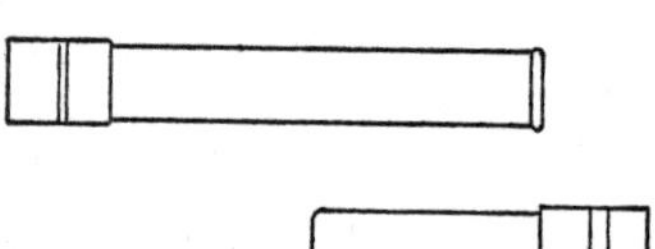

Fig. 4

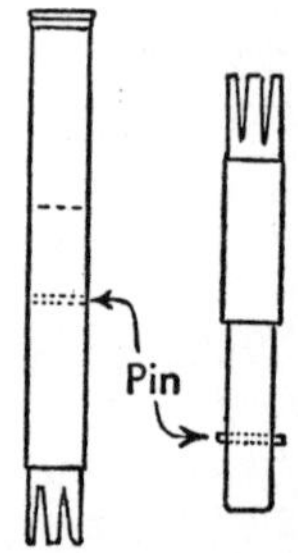

Fig. 7

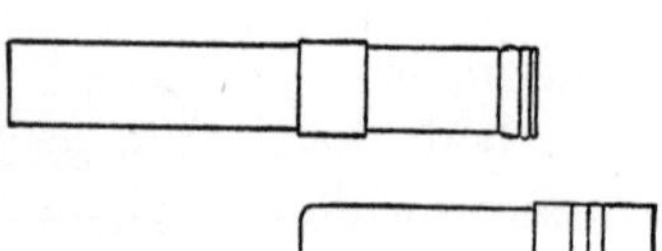

Fig. 5

for a light fly rod lower ferrule. It is of good length, the female member being 2⅝ inches long, thus having a bearing of 1¾ inches on the wood, sufficient for a good, tight seat. The male member, 2 inches in length, has a body an inch long, and a slide of equal length. This should be a minimum length of metal to metal bearing for a ferrule of this diameter, I believe, for it is only that 1-inch contact of metal which holds the rod together during severe use.

The open end of this female ferrule is heavily hand-welted with a strong ring of metal, hard-soldered to the tube. This welt, of course, adds strength at the weakest point and prevents distortion. These ferrules usually come with nicely tapered ends, not split or serrated. This work can be done at home with the little triangular hacksaw; or the manufacturers will do this work to order. You should specify whether for five- or six-strip work, and if a serrated or straight, split end is desired. I prefer the latter, as less metal is cut away, and the resulting narrow spaces between the five—or six—prongs of the end can be more neatly wrapped with the finishing thread.

Like all high-grade ferrules, these also have a metal disc fitted inside the female member at a point where the end of the

◀ FERRULES

1 *High-grade nickel silver ferrules, by the Orvis Rod Co.*
2 *The Peterson ferrule, with stainless steel female member, and the slide of copper beryllium.*
3 *Brass ferrules, nickel plated, of good design, but short length.*
4 *Heavy, brass ferrules, nickel plated, with shouldered ends.*
5 *Heavy-type ferrules for salt-water rods, the female strongly welted and with sleeve at center.*
6 *Ferrule driving plugs. (Chapter 15)*
7 *Showing where ferrules should be pinned. (Chapter 15)*

male slide meets it. This is important to prevent moisture from entering the wood where otherwise the end of the stick would be exposed.

Fig. 2 illustrates a pair of $^{15}/_{64}$-inch ferrules manufactured by G. C. Peterson, of 110 Bellwood Ave., St. Paul, Minnesota. These beautifully designed and finished fittings I believe result from Robert Crompton's search for the best possible ferrule which will fit snugly, provide long wear, with strength, and never seize.

The female member is of stainless steel, the male section, or slide, copper beryllium, a long-wearing, so-called "greasy" metal, which holds beautifully inside the steel without undue tightness, yet with a fine "glove" fit. This is a ferrule long for its diameter, insuring a good grip on the bamboo when properly mounted. The serrated ends are finely tapered and a waterproof disc is used.

There is also a new type ferrule lately on the market, the "Super Z," manufactured by the Super Z Manufacturing Company, Pearl River, New York. Made of 18 per cent nickel silver, it embodies a new principle, which seems of great value.

In this design the inside of the male ferrule is of constant size; that is, there is no "step down" from the barrel to the smaller size slide. Thus, the bamboo seat need only be shaped round, without reduction of the important outer fibres, to fit into the smaller part of the ferrule, as must be done in the conventional type. The result should be a rod much stronger at this critical point.

To fit this oversize male member the female of course must be larger. Thus it is swelled to the necessary size, but with a nicely tapered barrel and finely serrated ends.

While I have not yet used this ferrule, its striking advantages, coupled with its performance and appearance on rods I have used, together with a most beautiful fitting of the slides, convinces me that it is of outstanding value.

Other ferrules of good quality may also be on the market now, as more manufacturers recognize the need of high-grade rod fittings for the amateur workman. In the past, James Heddon's Sons, The South Bend Bait Company, and other large commercial rod builders have offered their ferrules on the retail market and may do so again. Both of these firms' products, in the best grades, are quite satisfactory.

In any case, when considering ferrules, do not hesitate to pay a substantial price, perhaps up to $2.50 a pair, for the top-quality article. Fig. 3 illustrates a cheaper, but fairly good, type of brass ferrule, nickel plated. These are to be found in sporting goods stores and supply houses, and they will serve, although lacking the strength and fine workmanship of the above mentioned articles. The design is good, and the wall thickness fairly heavy, as it must be when made of brass, a weak metal. The open ends are nicely tapered, but not split or serrated, and there is no waterproof disc in the female member.

Like most of the cheaper articles, this one is short, the slide in a 16/64-inch size being only 7/8 inch in length, and the female bearing only 1 3/8 inches on the wood. Neither dimension is enough to insure long service.

Then, there are various ferrules made of aluminum and aluminum alloys on the market. This material does not possess the necessary strength for the job, unless made with very heavy walls, and then, I believe, it has a tendency to crack under hard usage. It also seems to be a material difficult to keep on a rod, sometimes slipping even when a ferrule is carefully and tightly set.

Fig. 4 is a cheap type of brass, nickel-plated ferrules, often seen on the market, and generally used on the lowest priced machine-production rods. The illustration indicates the "shouldered" end of the female member, which has a short-beveled edge where it joins the wood. This type of ferrule, with its two diameters, seems to stay put when mounted, but its use involves taking off much of the best bamboo fibre, to fit the smaller portion of the barrel, or else the shoulder will be too large.

The choice in using this ferrule, then, is between the above bad practice, sacrificing bamboo, or, fitting the smaller portion to the diameter of the wood, when rounded, and then building up under the shouldered portion with thread, tape or plastic wood—all of which is rather bad practice.

Fig. 5 shows a heavy type of ferrule which is best for large rods undergoing continuous heavy strain, such as salt water rods. It is reinforced at the center, with a sleeve soldered in place, and usually has a long and heavy welted end. In nickel silver this is a splendid article; and those made of nickeled brass of heavy gauge may prove fairly satisfactory. This type of ferrule is needlessly heavy for use on fly and light bait casting rods.

Unfortunately, some manufacturers designate their ferrule sizes only by arbitrary numbers, and it is difficult to order what you need. It is best to know the exact calibre of each end of the sticks to be ferruled, then to transpose micrometer readings to fractions of an inch, and so order. Thus, a butt stick "miking," say, .295 to .300 would need a ferrule of 19/64 inch, or .297.

If the diameter went as great as .305, it would be best to use a ferrule of 20/64 inch, .312, rather than cut down the larger end to fit the .297 ferrule. This applies to six-strip work, a slightly different method being applied for the five-strip, as will be later explained.

FERRULING SIX-STRIP RODS.

If your workshop boasts a small machine or wood lathe the task will be comparatively easy and the work very accurately done. If you do not possess such a machine, perhaps a friend may lend you his facilities for an evening. The rod stick is inserted through the hole in the headstock, as shown in the illustration, and the rod padded with adhesive tape or thin strips of brass where the jaws of the chuck grasp it.

But first, you will have marked off each rod section, to be exactly the same length after the ferrules are fitted, and this takes a bit of figuring. It is best, perhaps, to lay the sections end to end on your bench between two marks indicating the entire rod length. Then lay each pair of ferrules at their proper place beside the sticks.

We are ferruling a three-piece, 9-foot fly rod for example, so each section, with ferrule mounted, will be about 36½ inches long, or a little more, depending on the length of the ferrules. Starting with the male ferrule on the middle section, lay it beside the wood, mark for length, and slide down to proper length beside the female. Mark at that spot on the butt section, then do the same for the tip ferrule.

Now, with a total rod length of 9 feet indicated, measure each section, with ferrule length included, and slide the ferrules when needed until all three sections match for length. Having started with 37-inch sticks for butt and middle sections, about a half inch will be cut from an end of each, to allow for the length of the male slide, which is something like an inch. This trimming may be taken from the small end of the stick, or a bit from each end if the ends are uneven.

Note here that the tip section, instead of being trimmed to allow for a female ferrule, loses almost an inch, or the length of its male slide. Thus, its over-all length will still be about 36½ inches, this being the reason the splines were first cut a bit longer than those for the butt and middle sections.

Trimming of the ends is best done with the little hacksaw, while the sticks are held in a padded vise or a lathe chuck. The saw should "go round," rather than straight through, to prevent splitting off the fibres at the finish of the cut. Finally, square the ends smoothly with a file. It is well now to check the length of the ferrule seat again, and measure *inside* the male member for its exact length.

When turning the ferrule seat in a machine lathe, use a tool with a long, very sharp point, and take light cuts, using fairly high speed. The end of the stick projecting back from the head stock can be steadied with one hand, or supported in a temporary jig, to prevent whipping.

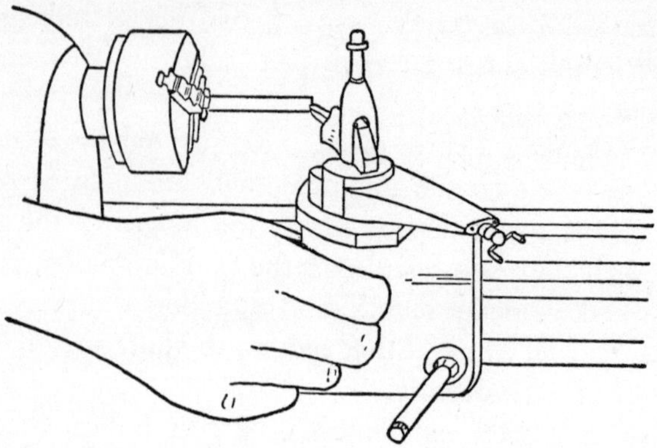

Using small lathe for turning down ferrule seat, the best and most accurate method.

The female ferrule seat will be exactly the size of its corresponding male member of course, so if this fitting is, say, 17/64 inch, the seat should be just that, or .266 inch. But use the micrometer frequently, and stop well before the last final cut is made to try the ferrule for fit. If strong pressure with your hand will just start it, discounting the spring of the serrated end, the seat is ready.

* The male ferrule will, of course, have two inside diameters, that inside the body, which is usually—but not always—the same as the outside diameter of the slide, and the smaller size inside the slide, usually .030 to .040 less. With two marks on the rod stick then, one indicating full length, the other the length up to the *inside* shoulder, start turning the end. When it is reduced to let the ferrule barely start, that will be the exact diameter for the larger portion to be turned later.

Now, turn down the end, stopping often to slip the ferrule up to the inside shoulder and "feel" for the first indication of its starting. There is no other way to determine the exact inside size of the slide, unless you happen to have a section of small rod or a drill shank the exact size, to serve as a gauge. Take an

* Except the "Super Z" or others of similar design.

hour, if necessary, to obtain a tight—and one-hundred per cent—fit.

If a wood-turning lathe is used for this work, without a tool holder, it is best to turn the seats with strips of fine sandpaper cut the exact width of the ferrule seat and held looped with the hands around the stick as it revolves. Very accurate work is possible in this manner, if the seat is reduced slowly and "miked" often.

In lathe fitting it is important to have the stick revolve accurately, of course. With a six-sided stick used in the usual three-jaw chuck, it may be necessary to try turning it a sixth-turn, since all the faces may not be exactly true. A strip of thin, waxed paper also will sometimes build up one side to give you a true job, which will insure a finished seat exactly concentric with the axis of the stick. A first thin cut, barely touching the corners, will indicate if the stick is ready for the deeper cuts.

The ideal result will be obtained when all the corners, barely reduced to a round seat, will provide the right diameter without taking off any wood from the flats other than to complete this true circular cut. Thus, all possible power fibres of the wood will be preserved. So important is this feature considered by some fine rod makers, notably Hardy Brothers in England, that extra bamboo strips are glued to the faces at each end, and these are later turned down to seat an oversize ferrule, thus preventing cutting into the wood of the actual stick at all.

But, you may sometime, despite careful planning, find your stick diameter a bit oversize for the ferrule diameter, necessitating the cutting of more wood for the seat, and causing loss of some of the outer power fibres. If this loss is not excessive, the matter will not be serious. However, there will be abrupt shoulders of the wood projecting above the ferrule end. In this case, remembering that the ferrule end is thin and serrated, so capable of expansion, taper that portion of the seat from the true diameter up to meet the wood "flats." This method will insure greater strength at this critical point and give the job a neat appearance, as well, when the serrations are later wound over with winding thread.

On the other hand, if the stick is a bit too small, you may turn down only the corners, leaving narrow flats, and fit the ferrule very tightly with the rounded corners only bearing on the wood. This method will prove quite satisfactory, particularly if you use the new cold type of ferrule cement. Another method which will serve, if your stick is badly undersized, is to turn down a round seat, then build it up with rod wrapping silk to the required diameter. Thus, "00" silk will increase the diameter of the seat about .008 of an inch, and the thread will provide a good holding contact for the metal.

The tail stock of a lathe will provide adequate pressure in forcing the ferrule into place. Now, if you use the hot type of ferrule cement, which can be purchased in any sporting goods store in small sticks, a small Bunsen burner or alcohol lamp will be useful. A candle, however, will serve, though it is smoky and something of a nuisance. The cement is softened by holding the end of the stick over the flame, and then is applied to the wood with a knife while the lathe is turned slowly by hand. A thin film only is required. Now, run up the tail stock, clamp it at the right spot, warm the ferrule and wood, and start the ferrule by hand. Warm it again, place a block of soft wood between its end and the tail stock for protection, and slowly apply the pressure. Several heatings may be necessary before the ferrule is fully seated, if the seat has been turned to a full size.

But excessive pressure may distort the metal, an indication that the seat is perhaps one or two thousandths too large. The principal advantage in using the usual type of hot cement is that the ferrule can be removed at any time by application of heat to resoften the thin film under the metal. But, there should be no reason to remove it, granting close fitting, well-seasoned wood which will not shrink, and a good ferrule.

A comparatively new type of cold cement I have recently tried seems far better than the stick cement applied hot, and should replace it for rod building work, since a tight-fitting, lifetime job is so essential. This cement is sold generally for household use, for joining such difficult materials as glass and leather. Several varieties, apparently much similar, are sold under various trade names, and I have found the Goodyear product, "Pliobond 20" to be excellent for ferrule work.

This material, of heavy gravy consistency is applied to the ferrule seat with a brush or knife blade, permitted to become quite tacky, and then is ready for the press fit of the ferrule. Both metal and wood must be clean for best results, and the ferrule should not be used or manipulated for several days, or until the cement has thoroughly hardened. A wait of two weeks or more before the rod is used seems to let this material harden and set until a ferrule can only be removed by destroying and chiseling it off. My tests with this new adhesive, to date, over a period of about six months, seem to indicate that it will not powder, dry out, or disintegrate, as the hot cement sometimes will.

A final caution in pressing on the female ferrule: Be careful never to force it up on the stick even a fraction of an inch beyond the carefully marked seat, or the male slide will not seat its full and proper length.

FERRULE SIZES IN 64ths OF AN INCH, AND DECIMAL EQUIVALENTS IN "THOUSANDTHS."

NOTE: Beyond a size 7/64th inch the final decimal is eliminated for convenience, and the nearest fraction given for a three-figure reading.

64ths	*Thousandths*	*64ths*	*Thousandths*
1	.0156	14	.219
2	.0312	15	.234
3	.0469	16	.250
4	.0625	17	.266
5	.0781	18	.281
6	.0938	19	.297
7	.1094	20	.312
8	.125	21	.328
9	.141	22	.344
10	.156	23	.359
11	.172	24	.375
12	.188	25	.391
13	.203	26	.406

CHAPTER 14

HAND FITTING FERRULES—FINAL LINING UP AND DEFLECTION TESTS

The ferrules can be accurately and tightly fitted by hand if sufficient care is used. But this cannot be accomplished by the usual crude method of using a pocket knife and rasp. First, hold the end of the stick in a "V" block or the partly opened jaws of your vise, while you revolve it and score a shallow knife cut around the pencil marks.

Then clamp the stick in the padded vise, with the mark flush with the jaws and the end protruding, and use sandpaper strips drawn around the upper corners with long, easy strokes. After a few strokes, turn the stick one flat in the vise, repeat with the same amount of rubbing, and so on around. By thus turning the stick at regular intervals, the seat will be kept concentric and also round. Use the micrometer often, at all points of the seat, for there is a tendency to take more wood from the end than at the shoulder. This is the reason for the light score mark there, to permit the edge of the sandpaper to bear sufficiently hard at this point.

When near to size use the micrometer again, finally rubbing lightly any spot which is a trifle high, and so bring the seat down to the exact thousandth required. You may easily spend an hour shaping one seat properly, and the male section will need special care. But the effort to secure a perfect fitting job will be decidedly worth while. After applying the cement, either the hot or cold type, the ferrule may be pressed on by holding the stick at exact right angles and close up to the edge of the bench or a block of wood set in the vise. Steady, even pressure is

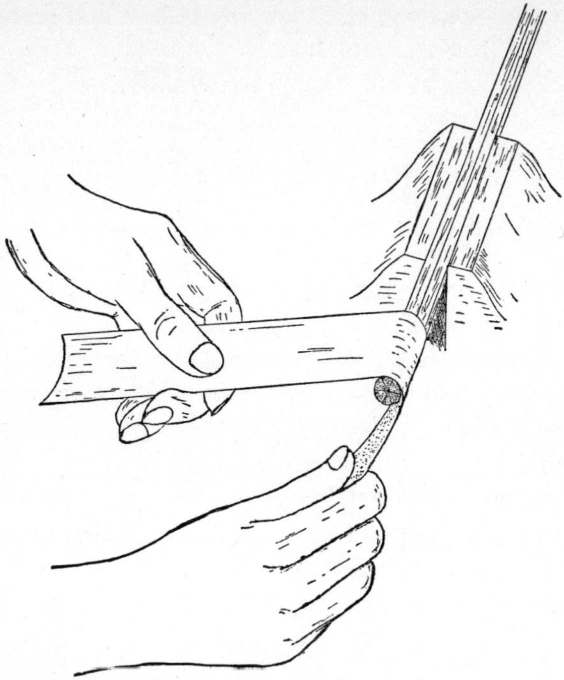

Shaping ferrule seat with sandpaper strip, rod held in vise.

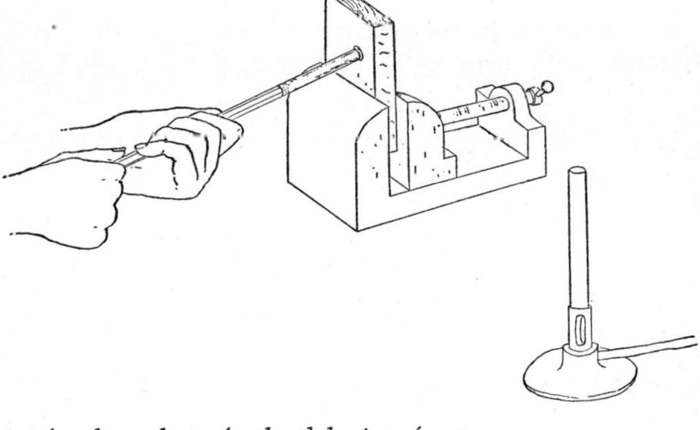

Forcing ferrule to place, after hand-shaping of seat.

necessary, and several mild applications of heat may be necessary to fully seat it. Be careful here, for distortion of the stick is possible, especially if excessive pressure is used when ferruling slender tips.

SETTING FERRULES WITH DRIVING PLUGS.

Fine results can be secured with a driving fit, but this of course cannot be done on the bare ferrule end. These plugs are very worth while for the purpose, especially if you expect to build several rods using similar size ferrules. They can easily be made in your own lathe, or by a machinist at little cost. Mild steel is suitable material, and brass or duralumin will serve nicely, being easy to work.

The plug for the female ferrule should be an exact "slip-fit," and the length of the male member which, in fact, it replaces for this driving operation. The square shoulder, supported by the bearing of the walls, distributes the strain evenly on the welted end of the ferrule as it is driven home with smart taps of a mallet.

The plug for the male section likewise must be a close fit to the slide and have an inside length which will touch the ferrule end, as well as bear strongly on the shoulder. After starting the ferrule by hand it is warmed again, the plug applied and tapped firmly. Several repetitions of warming and tapping may be necessary, if the seat is truly tight. This method will produce a fine, straight-line job, and permit very tight fitting. The rod stick, however, should be held in the other hand, close up to the ferrule, and not rested against a solid support.

These plugs, a pair of which are shown in the illustration, Fig. 6, are also useful when pressing on the ferrules with the lathe tail stock, for then great pressure is possible without harming them in any way.

PINNING FERRULES.

To pin, or not to pin is a question which seems difficult to answer. Some of the finest custom-made rods have the ferrules neatly pinned, and so nicely finished that even with a magnifying glass this cannot be ascertained. Many "production rods" have pinned ferrules, also, and give satisfactory service. Again, the makers of other fine rods frown at the practice.

My own humble analysis of the problem is that with well-seasoned sticks, no longer liable to shrinkage, and with very tightly fitted ferrules, we might "get by" with years of service without a sign of looseness. This is true only in a locality having average humidity. In the bone-dry air in many parts of the West, bamboo still can—and will—shrink that bare thousandth or so which will let an unpinned ferrule loosen. I know by sad experience.

So, for a "lifetime" job, I believe that *proper* pinning is desirable, particularly on fly rods, which with their great flexion seem to just want to fling the ferrules off. This, of course, does not actually occur, but the whipping action does sometimes start even well-fitted ones loose from their seats, unless pins are used.

Nickel silver wire is best for the purpose and may be obtained from a metal supply house and, sometimes, from hardware stores. Buy a size .040 inch in diameter for the large ferrules and size .032 for tip ferrules. Twist drills No. 62, diameter .037, and No. 68, measuring .031, respectively, will give you the proper size holes for a neat drive fit of the little pins. Cut these a little longer than the ferrule diameter, and sharpen one end a bit with a fine file, for easy starting.

A drill press is best, of course, for drilling the holes, with the ferrule resting in a "V" block. However, the ferrule may be clamped in the padded vise and the work done with a small hand drill. The female member should be drilled about 3/8 inch below the end of the stick—not down near the serrated end, as is sometimes done. The male slide should be drilled at about the center (note the illustrations, Fig. 7).

It is a bit of a trick to start these small drills without their "running," and you may need to rig a steady-rest to help you. Keep the drill at exact right angles to the rod stick, and strive to get the hole centered. It should start to enter at the center of a flat of the stick, and emerge exactly opposite, on the opposing flat.

Rest the ferrule on a block of wood and tap in the pins smartly, with a small, light hammer, until the ends emerge. Snip off all the surplus possible with cutting pliers, then finish the ends flush with a small jeweler's or Swiss file, having a narrow, but flat, edge. File carefully, with a light, circular motion, and the end of the pin will merge with the ferrule metal, rounded, smooth and flush. It *must* be flush in the male slide, or that pair cannot be assembled. Finally polish each pin end with a very narrow strip of the finest emery paper.

FERRULES FOR THE FIVE-STRIP ROD.

Here the work is essentially the same as for the "six," with one major exception. More corner will be removed from the five-strip to secure a round seat, as previously noted, so a ferrule size about $\frac{1}{64}$ inch less than for a six-strip will be right for the "five," having the same over-all calibration. But again, take off no more wood than necessary to *barely* provide the seat. The corners at the end of the ferrule may be tapered down with a strip of sandpaper to be flush with the outside of the serrated end, for a neat job of wrapping later.

When using a lathe for five-strip ferruling, the pentagon will not fit the chuck of course, since it has three jaws. But it is easy to make a little split bushing from a piece of light tubing, or an inch cut from an old ferrule; slit one side of this with the little hacksaw, and slip it over the wood where it is clamped by the chuck. Thus, the five corners of the stick will bear equally on this bushing, while its circular shape will center in the chuck. The slit permits it to spring enough to take up any slack.

Ferrule pinning is the same as for six-strip work, except that the pin is started in line with a flat and comes out through the opposite apex.

The final step now, before the sticks are ready for the fittings, is to line up the ferrules. This is important. Rarely can they be so perfectly aligned with the axis of the sticks that a haphazard assembling of the rod will find them all in a true plane. So start with the butt and middle section jointed, sight down the length and if necessary turn one section to the next flat, and sight again. Thus there will be six possible adjustments—or five, in the hexagonal rod—one or more of which will permit perfect alignment, if the ferrules were carefully centered when seated.

Mark the aligning flats with a pencil, close up to the ferrule end, then fit and mark in similar fashion the tip sections. Now, if your sticks are perfectly straight the rod should deflect about the same as you hold it horizontal toward a light and slowly revolve it. In this case any side may be marked to indicate the mounting of guides and reel seat.

But there may be slightly more deflection when one particular flat is held uppermost, since we can hardly hope for absolute perfection in this first attempt at rod building. Then, if this is a fly rod, just mark for the guides the flat showing least deflection. In later use the strain will tend to oppose and so correct this slight imperfection. In bait casting rods, with reel and guides used on top of the rod, the matter is reversed, and the guides should be placed on the side which deflects most.

Stiffer rods of the salt water type will not show any change in free deflection and will require a weight of several ounces hung to the tip for a conclusive test. This also may be done with fly rods and will permit an interesting study in rod arc, the result to be noted on your rod chart and kept for future reference. Other completed rods, perhaps one you may wish to copy for its fine action, may be studied in similar fashion.

Such a test will indicate many things we can only guess at while using the rod: just how much tip action is developed, and how far it extends down the rod; where the power of the middle section comes into play; and how much a moderate working arc will affect the butt section.

Clamp the rod butt in a padded vise, the tip extending horizontally across the room, then stop its vibrations and measure the distance from the end of the tip to the floor. Then hang a two-ounce weight to the tip and measure the amount of deflection. In fly rods this will range somewhere from 17 to perhaps 25 inches, depending upon the rod length and calibre. You may go farther, too, in noting deflection for future reference, and make a complete picture of the rod under strain by measuring the arc at each foot of its length, and retaining these notes along with the calibration chart of that particular rod.

You may have heard at some time of a vibration test for light rods, in which the rod butt is clamped tightly, extended horizontally, and made to vibrate violently by pulling back and releasing the tip. A count of the vibrations is made for a few seconds at the time of highest intensity, then multiplied to indicate the movements per minute. This test is interesting, but seems to prove little. It is sufficient, I think, simply to observe how quickly the tip stops its movement. A quick stop, without weave and wobble, will at least prove the rod to possess well-designed power and that important requisite that anglers call "backbone."

CHAPTER 15

REEL SEATS AND HAND GRASPS

Reel seats are almost as varied in design and weight as are rods, and a wide variety of material is used in their construction. In the older rods of high grade these fittings were usually made of the nickel alloy called German silver. It is strong, machines beautifully, and lends itself well to the "drawing" processes sometimes used in shaping it. This material, while still being used for heavy-duty reel seats on some fine salt water rods, is apparently difficult for manufacturers to procure. Thus brass, heavily nickel- or chrome-plated, is now in common use. It is not as strong as nickel; bearing points, such as screw locking devices, wear faster, and it is heavier.

However, weight is of comparatively little account in the heavier type of rods, while the secure anchoring of the reel, with a positive locking device, is of paramount importance.

Figure 1 illustrates such a fitting of brass, heavily chrome-plated, suitable for light salt water and heavy-duty trolling rods. Its inside diameter is ¾ inch, and it is 5 inches in length. A double locking ring screws up against the sliding hood, inside of which is a smaller hood which is split, to provide a "jamming" effect, and so secure the tail-piece, or foot of the reel, very tightly.

FLY ROD REEL SEATS.

Here, light weight is of considerable importance, although we need not sacrifice a mere fraction of an ounce which will help hold the reel securely just to keep the rod's over-all weight down to a minimum. To attempt to define a rod's action and potentialities by weight alone is impossible, as discussed elsewhere, so it is most unwise to try for extreme lightness in over-all weight when impractical fittings are used.

It *is* good practice to use the lightest *efficient* type of reel seat, particularly on the "fairy" fly rods of light calibre, for such tools cast better when unhampered by unnecessary weight of fittings. But heavy-duty rods, for salmon, bass and steelhead, require a positive-locking type of seat for the larger reels used for the heavier work.

Figure 2 illustrates a screw-type seat of duralumin, usually found on the present-day market, and is of excellent design. The barrel is heavily treated with lacquer and comes in a variety of colors. It has a scant diameter of ¾ inch inside and is 3½ inches

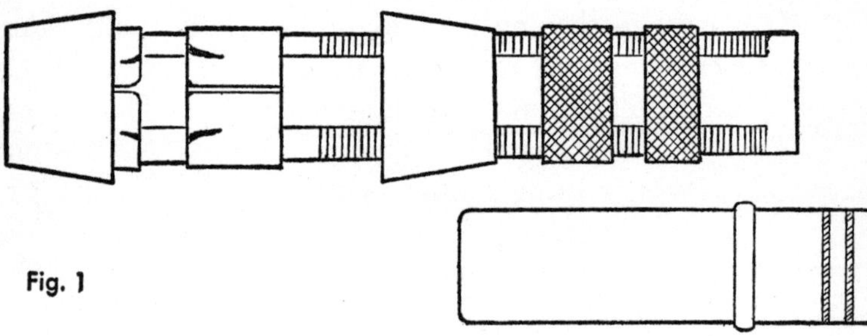

Fig. 1

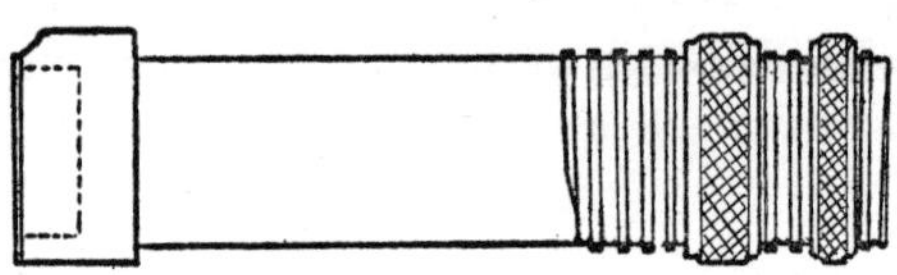

Fig. 2

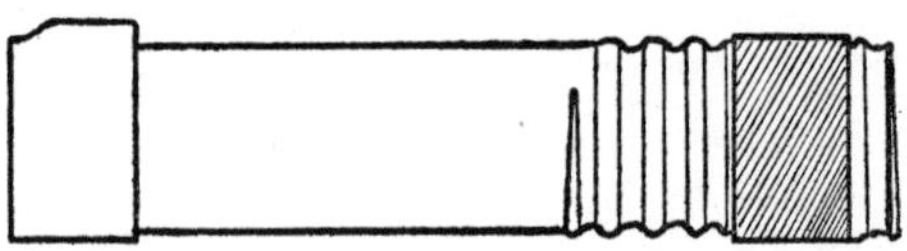

Fig. 3

REEL SEATS

1 *Heavy-duty reel seat of brass, chrome plated, for salt-water rods, with ferrule.*
2 *Duralumin fly rod steel seat, with double locking ring.*
3 *Similar type, with single, threaded ring.*

long. The hood is strongly pressed on, and the end cap, pressed into the barrel, may be removed when the seat is to be mounted on a bait casting rod, with the reel above the hand.

The main locking ring of this seat is undercut, for a good fit over the reel foot, and is held tightly in place by the jam ring.

Figure 3 shows a fitting of similar size and material, having a single, and wider, locking ring. Its weight is a bare ⅝ ounce, as against the ¾ ounce for the seat having the extra ring.

Figure 4 illustrates a somewhat different design of fly rod reel seat, made in both ⅝ and ¾ inch inside diameters, employing a sliding hood, held tightly in place by the knurled, threaded jam ring.

Fig. 4

Fig. 5

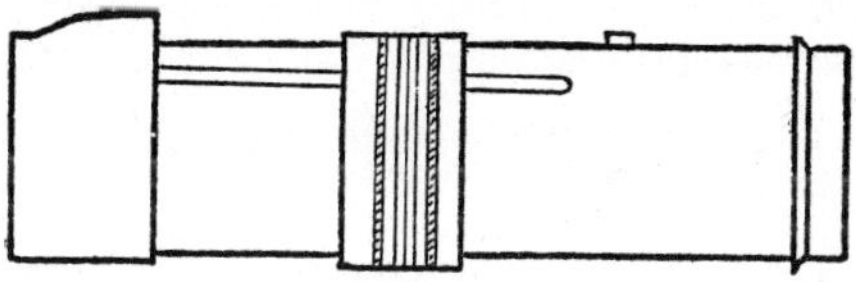

Fig. 6

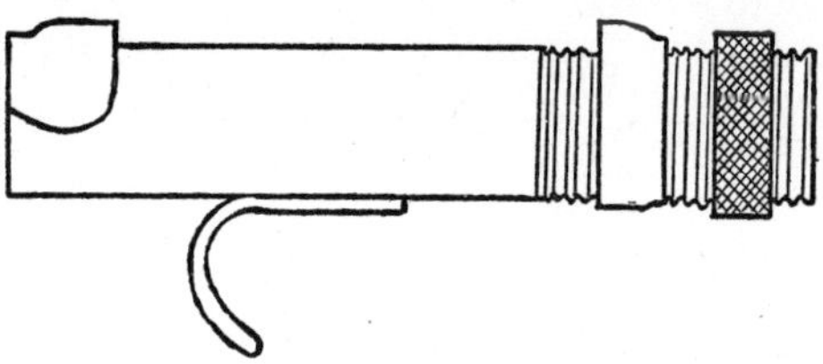

REEL SEATS

4 Light, metal fly rod reel seat. ⅝ inch outside diameter.
5 Nickel fly rod reel seat, with sliding ring.
6 Good type of nickel reel seat for the bait-casting rod, with sliding hood and threaded jam ring.

In Figure 5 we have the older style nickel seat, with only a sliding ring to secure the reel. Two ridges, expanded from the metal, are intended to prevent lateral slipping, but even then this seat is unreliable for heavy work unless the ring can be jammed up very tightly over the foot of the reel.

There are other very nice seats made with plastic barrels of various colors, having metal hoods and screw-locking rings which are threaded on the barrel. You perhaps have tried and liked some of these, found on the best commercial and custom-built rods. But they seem hard to procure, and are probably built specially for the rod manufacturers.

REEL SEATS FOR BAIT-CASTING RODS.

Figure 6 illustrates a nice seat, designed for bait casting rods, and having a finger hook, which most anglers prefer. This is all nickel, with a soldered-on fixed hood and a sliding hooded ring, held snug by a threaded jam ring. Like all bait-casting seats it of course mounts ahead of the grip, so is open at both ends.

The fly rod reel seats, Figs. 2 and 3, are also nicely adaptable for the bait rod, since their end caps can be knocked out without disturbing the hood. Also, their size, ¾ inch, inside, is proper for this type of rod. If a finger hook is desired it may be fashioned from a strip of malleable metal and riveted or screwed to the barrel of the seat at the proper place.

All the above reel seats are mounted with a soft wood filler, usually available with them. This fits snugly inside the barrel and usually has a hole about ⅜ inch into which the bamboo is fitted and glued.

Finally, the lightest type of seat—though not always the most practical—is one having a barrel of cork or fine-grained hardwood, with a hooded cap fitted to the end, and a sliding ring to hold the reel. Using cork, we will have the lightest possible means of holding the reel in place, and this type of seat may be desirable on the tiny fly rod of 7 or 7½ feet, where weight is an important factor.

Curly walnut, birch, maple or myrtle wood all are suitable for making a handsome barrel for use with a sliding ring. These and other hardwoods will take a fine polish and prove durable. But always, when using this type of fitting, the wood must be shaped to a size which will barely let the ring, which should be thin enough to be capable of mild distortion, to jam *very* tight over the reel's foot; otherwise you will have annoying trouble when you go fishing.

While the sliding-ring seat is best adapted to fly rods it may also be used on the light bait rod when the reel is carefully fitted to it, so there is no wabble or lateral motion.

HAND GRASPS.

Cork grasps are almost universal and they are satisfactory on all light types of rods. The cork is light, provides a good, non-slip grip, and is easy to apply and shape. Cork rings, made for rod use, come in various grades and sizes and often may be had at local sporting goods stores. However, rings having many and deep holes or flaws will not produce a nice job. I have had the finest grade, designated as "3X, Spanish," from the firm of Edward Vom Hofe and Company, 113 S. 16th St., Philadelphia.

These rings come ⅜ inch thick, 1⅜ inches in diameter and may be had with holes from ¼ to ½ inch, to fit various sized sticks. The cork will stretch when shoved into place, so it is best to order the rings with holes smaller than your stick diameter. Ready-shaped grasps of glue-up rings are also to be had, but it is difficult to fit this assembly tightly to the rod; the material is often of poor quality, and the shape of the grip may not suit you. For these reasons, building your grip of the rings is probably best.

Here again, if you have a small lathe, shaping the grasp and fitting the reel seat—the two operations being somewhat connected—will be simplified.

First, mark off clearly the length of the reel seat on the butt stick, then for the length of grasp you desire. Now, if for a fly rod, fit the wood filler tightly in place, sanding off the corners of the stick down to make a circular seat, but no more. Otherwise, drill out the filler to proper size. If the filler hole is too large, the stick corners can be rounded, then wrapped closely with cotton string to obtain a good fit. The filler then is cemented and pressed to the mark. I have found either Cascamite or Weldwood powdered glue excellent for both this work and gluing the cork.

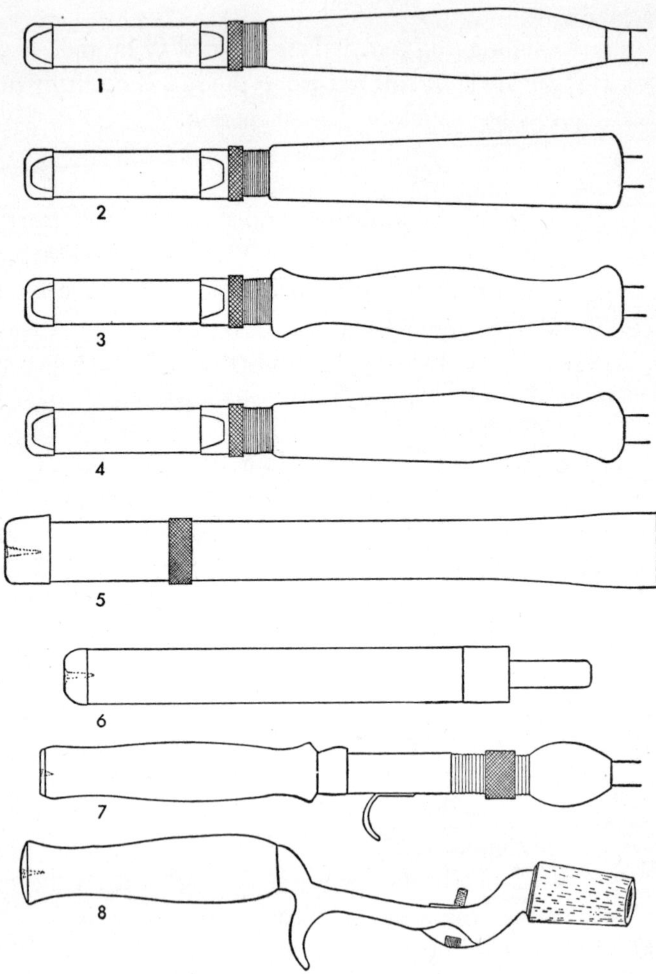

These are absolutely waterproof, and may be mixed thick for best results.

If you are building a bait casting rod, it is best to fit the reel seat filler in place, then remove it and glue up the cork rings for the main hand grasp; then fit and cement the reel seat in place, and finally add the rings for a forward grip, if it is desired.

Be sure the rings will fit the stick snugly as they are forced from the small end down to place. Smear on a good layer of glue to each face and press firmly against the side of your vise, with the butt of the stick inserted loosely between the jaws. About sixteen rings of ⅜ inch size will build a fly rod grasp of the conventional 6-inch length, but you can, of course, follow your own desires both as to length and shape.

With a lathe having a hollow headstock you need only insert the stick with the assembled rings toward the tailstock, use a large washer at each end for even pressure, and screw up on the tailstock firmly. All surplus glue will thus be forced out, and the rings pressed so tightly together the joints will not be visible. The work should be left here about twenty-four hours for solid setting of the glue. To shape the grasp then you need only support the extended ferrule end of the stick, insert a center in the tail-

◀ HAND GRASPS FOR FLY AND BAIT CASTING RODS

1-2-3-4 *Various shapes of cork fly rod grasps. No. 3 is hand-fitting and suits a majority of anglers.*
5 *Extra long grasp, often an advantage on heavy duty fly rods.*
6 *Detachable cork grip for heavy fly rods, the extension permitting two-handed work. Ferrule fits socket under reel seat.*
7 *Good type grasp and reel seat for one-piece bait casting rod, the bamboo extending through to butt.*
8 *Off-set, detachable handle for one-piece bait casting rod, socket taking a ⅜ inch ferrule.*

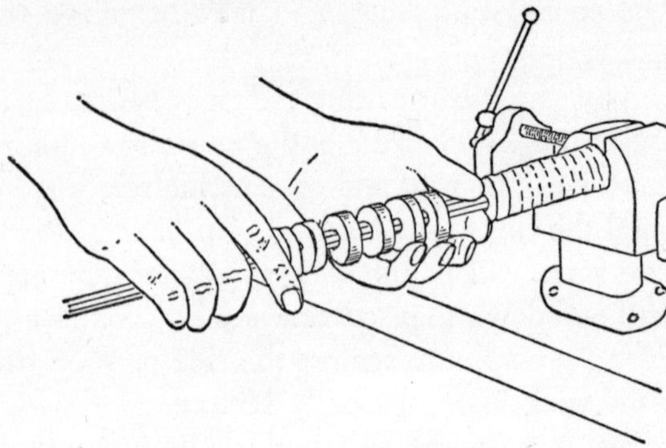

Cork rings for hand grasp being pressed to place in vise. A thin coating of resin glue is applied to each face and the rings fit the stick tightly.

stock, to bear on the wood filler, and turn the cork down to shape with sandpaper strips.

Fairly high speed works best, and much of the surplus cork can be taken off with coarse-grit paper. A number of popular style grasps are shown in the illustration, most of them having a largest diameter of an inch, or a bit more, so the corks may safely be reduced to this size, before final shaping is done. Then, it is best to use No. 00 paper, cut in narrow strips which are continually moved back and forth to prevent channeling. If you are not sure of the exact shape desired, stop and try it for "feel" as the work progresses. It is nice to finish the grasp finally with

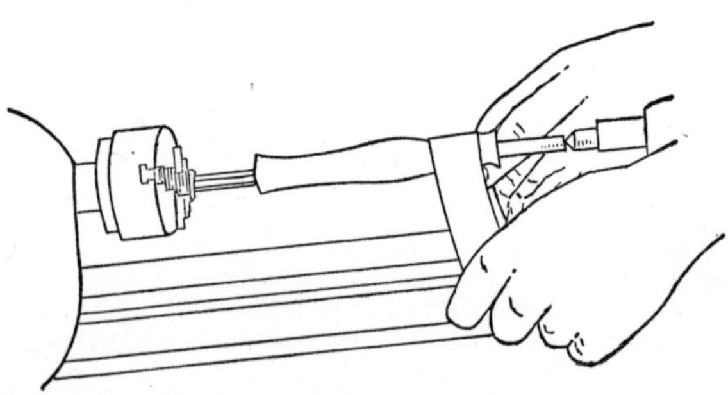

Shaping cork grip in lathe with sandpaper strip.

a strip of No. 280 "Trimite" paper, for a fine, velvety surface.

The final operation now is to fit the reel seat, exactly as you would a ferrule in the lathe: Assuming it is metal, apply hot ferrule cement to the wood filler, warm the seat, line it up carefully—this is easily forgotten at the moment—and press it home with the tailstock.

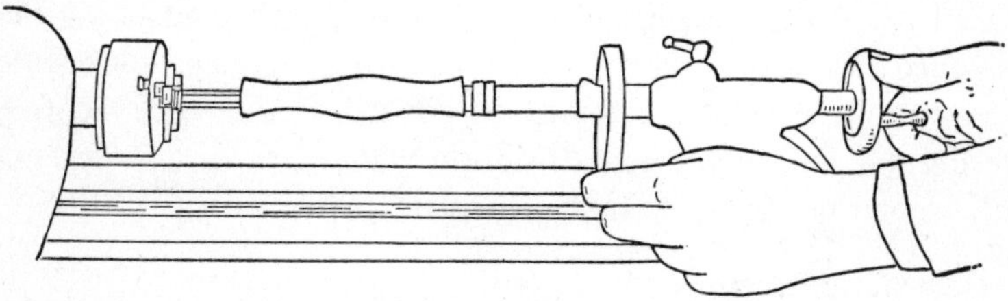

Pressing on a reel seat in the lathe, with pressure from the tailstock.

If your reel seat is to be cork, the rings will, of course, have extended to the butt of the stick, and the seat turned down to the required diameter when the grasp is shaped. In this instance a shallow center hole will have to be drilled into the butt of the stick itself for insertion of the "dead center" in the tailstock, to hold the work while it is spinning.

If a hardwood barrel with sliding ring is used, it will have been fitted and cemented into place instead of the soft wood filler. But enough surplus should have been left for final shaping and polishing in the lathe after the hand grasp is finished. Turn down carefully the shallow seat for the hooded cap, which should go on with a tight "press" fit, its outside exactly flush with the wood of the barrel.

For a fine finish the hardwood should be gone over lightly first with No. 00 sandpaper, then with No. 280 Trimite paper and, finally, with No. 00 steel wool and pumice powder, until it is perfectly smooth. But allow surplus wood for this polishing, and be sure that too much is not taken off to spoil the tight fit of your reel when held with the jam ring. For final treatment I like to wet a rag with thin shellac and hold this on the spinning wood until it becomes hot and "burns in." Successive applications will fill the grain of the wood and so repel moisture, as well as produce a fine finish. This can be enhanced by applying linseed oil with the rag, again, until the wood becomes hot. But, if you prefer a varnish finish, just apply the shellac as a filler, and then varnish the reel seat later when the entire rod is treated.

Do not forget to slip on the sliding ring before attaching the hooded cap, which may be set either with the glue or hot ferrule cement. In any case, it is wise to pin it for additional security, using the .040 nickel wire driven through a hole drilled through the center of the cap to clear the hood.

HAND-SHAPING THE CORK GRASP.

This will prove a bit more work than when it is done in a lathe, but an equally good result is possible if you are willing to devote perhaps an evening to this important phase of the work.

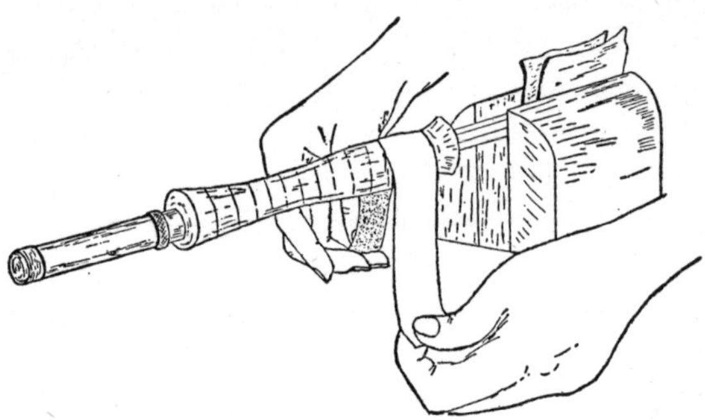

Shaping cork grip by hand, rod held in vise.

I have found it easiest to take off much of the surplus cork by revolving it slowly in the partly opened jaws of the vise, while a medium-fine wood rasp is used with both a round and lateral motion. Well before the minimum size of the grip is reached, clamp the stick in the padded vise jaws, and start working around the grasp with a strip of sandpaper, just as in fitting ferrules by hand. Then gradually apply the finer grades, as recommended for lathe work.

It is wise, also, as in ferrule fitting, to turn the stick from flat to flat in the vise to keep the grip concentric. If the bamboo springs and so makes the work unsteady, try placing the butt end of the stick against your belt buckle, as you stand close up, facing the vise. Such simple solutions of many small problems will present themselves in every phase of this interesting rod work, and you will be amazed at your own inventive ability. Already, perhaps, you will have discovered new and better ways of doing certain things than I have been able to describe.

The final type of cork grasp yet to be discussed, of those in general use, is for the spinning rod. Here the grasp *is also* the reel seat. The grasp is from 12 to 14 inches long, of about 1$\frac{1}{16}$ inches diameter, and the two sliding rings, of nickel or duralumin, permit the reel to be mounted at any point below the hand, for the balance and casting action desired. The corks are applied and glued in the same manner as described above, then should be shaped to let the rings barely slide along the length of the grasp. Most spinning reels have nicely tapered feet, which will let the rings jam snugly over them and, with this force, press the metal into the yielding cork for secure holding. Such security is essential for the spinning reel, whose center of gravity is well above the rod.

There is something to be said for this type of "continuous" grasp on a fly rod, too. It permits shifting of hand and reel to give a somewhat different balance of the rod and, in some rods, a decidedly different and even better action. Long, heavy salmon and steelhead fly rods may be used with less wrist fatigue with such a grasp, with the hand slipped well forward, permitting the butt of the rod to rest under the arm for support. This helps particularly when playing a heavy fish.

This long grasp, regardless of the reel mounting, may be well employed on any heavy duty rod, and for the same reasons as mentioned above. I once used a steelhead rod having a 14-inch grasp, above the reel seat, and found it very comfortable to use a two-hand hold on the rod when a heavy fish was fighting long and "going places."

A detachable butt is another way of having this advantage on the heavy rod. This is used on many of the fine rods built for salmon fly fishing. It may be carried in a pocket when not in use, then quickly slipped into the butt of the rod, providing a comfortable hold for one hand below the reel, the other above.

To build this, insert a large-size female ferrule, perhaps ⅜ inch, inside the wood filler of the reel seat, the open end to be flush with the end of the hood. The end of a one-piece hood will have to be sawed off, or the end plug driven from a reel seat built in that manner. The ferrule then should be accurately fitted to the end of the butt stick, and to the proper length, before the reel seat is permanently placed.

Use a hardwood dowel ½ inch in diameter upon which to build the detachable butt, fitting the male ferrule to it first, then slipping on the cork rings, to be glued and shaped as you did the rod grip, to any desired size and contour. Ordinarily a straight grasp for this butt seems best, and a diameter of about an inch seems to fit the hand. The length may be from 6 to 8 inches.

CHAPTER 16

GUIDES AND TIP TOPS

It is important to choose the proper guides for this new rod, from the standpoint of efficiency in its use, the prevention of unnecessary wear upon the line and long service from the guides themselves. Guides of soft material, such as those of brass, thinly nickel-plated, often noted on cheap rods, will quickly "channel," develop sharp shoulders at the edge of these little depressions and soon ruin a fine line. Replacement then is laborious and expensive. Certainly this new rod, over which you have labored so long and diligently, will deserve the very best fittings.

FLY ROD GUIDES.

The friction set up by the line continually "shooting" through the fly rod guides is considerable, especially since a line of large size in comparison to guide diameter is used. The line, of course, must match or "fit" the rod's action if the tool is to perform properly, so we can only choose a happy medium in the size of the guides which will help reduce friction and at the same time keep down the over-all weight added to the stick. Unfortunately, most guides of the ring type, while producing little friction, are too heavy for use on the smaller portions of a fly rod. Five or six of these mounted on a slender tip section would obviously affect its action and result in the entire rod feeling slow and logy.

The conventional "snake" guide of tungsten steel is used almost universally on fly rods, except for the butt or stripping guide, and serves well, although subject to some wear at the shoulders where the line bears a bit as it enters the circular portion. This type is probably the lightest readily available and may be found in almost every tackle store, in sizes from the very large, No. 3, down to No. 4-0, suitable for the finest rod tips. They are colored a dark blue or black, so are inconspicuous on the rod.

Snake guides similar in size and shape are also made of very hard, fine wire which is chromium plated to resist wear and rust, and these will serve a lifetime without channeling or developing sharp shoulders. The wire in these guides, however, will bend rather easily and may become distorted with careless use, necessitating sometimes careful straightening with round-nose pliers. These guides are attractive on a rod of light color, with silk wrappings of a light shade, and are decidedly practical if you are not averse to the shine of the polished chrome. Both of these types of guides are of the well-known "Perfection" brand, I believe, and may be ordered with assurance of fine quality at a very moderate price.

Other fine snake guides are sometimes offered in phosphor bronze, an extremely hard metal. These are usually of British manufacture and will often be found on the fine rods built there and exported to this country. With the constant improvements in metal for rod fittings and the adaptation of new designs for rods of various types, we can expect even better guides in the near future. Already, a new, very light and hard ring guide has been developed for the increasingly popular spinning rod.

The fly rod butt guide must be comparatively large since it first receives the line whipping from the reel and so straightens it out more or less before it reaches the other guides. Most fly rods, then, except the very light ones, should have a butt or stripping guide with an inside ring diameter of at least $^{14}/_{64}$ inch. Heavy fly rods will stand a $^{16}/_{64}$ inch size, while the $^{12}/_{64}$ is about the smallest practical size for use on even the "fairy" 7-footer. Here again the Perfection, chrome-plated guide, seems the best.

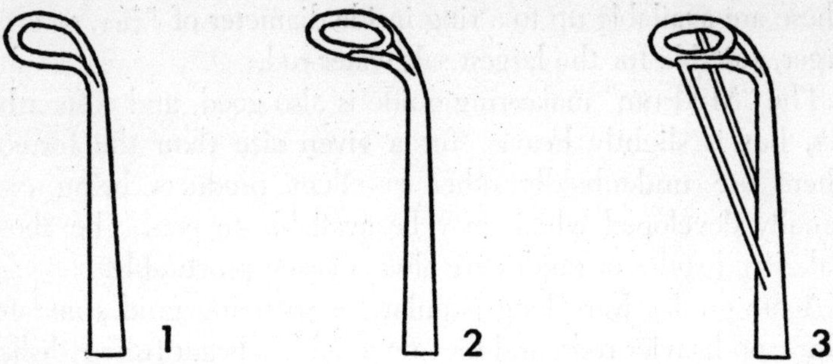

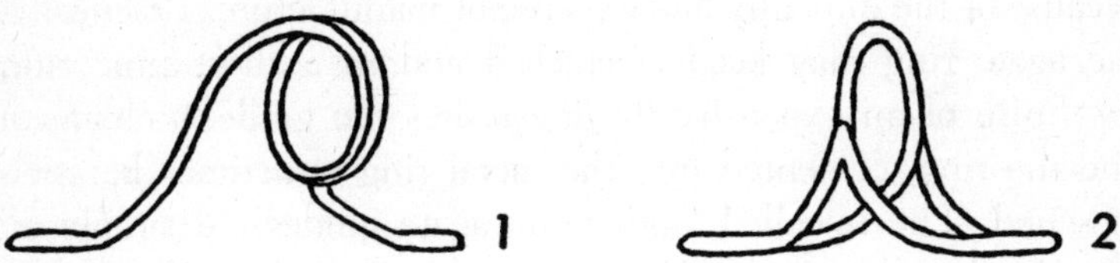

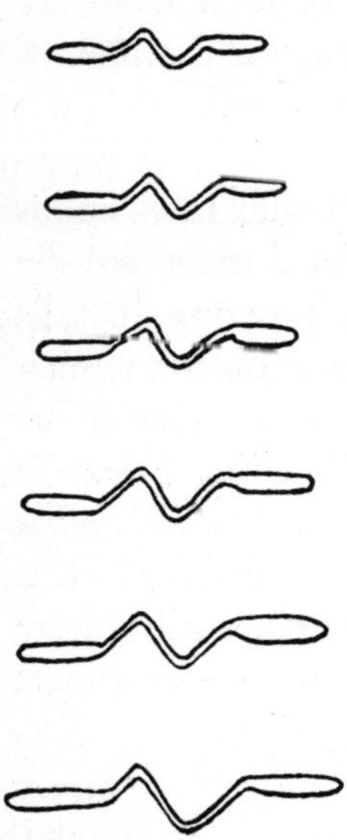

ROD FITTINGS

TOP

1 *"Pear-shaped" fly rod tip top.*
2 *Ring type, fly or bait-casting tip top.*
3 *Heavy-duty bait-casting tip top, with struts.*

MIDDLE

1 *Berner, monel, chrome plated guide, for all heavy duty rods.*
2 *Perfection ring guide, chrome plated.*

LEFT

Tungsten steel fly rod snake guides, showing comparative size, from No. 3-0, smaller, to No. 3.

These are available up to a ring inside diameter of $\frac{32}{64}$, or even larger, suitable for the largest salt water rod.

The "Mildrum" make ring guide is also good, and costs a bit less, but is slightly heavier for a given size than the former. There are undoubtedly other excellent products being continually developed which may be available to you. The above makes and styles of guides are always easily procurable.

Agate guides were long popular for stripping guides and for all use on heavier rods, and they do provide a beautifully polished bearing surface as well as a handsome appearance. But they have been almost entirely discarded on American-made rods, possibly because of the difficulty and expense of manufacture. Practically, the agate ring may become cracked and in a short time ruin the finish of an expensive fly line unless the guide is changed; and the ring, cemented into the metal ring sometimes becomes loosened. The so-called "imitation" agate guides are simply an abomination. The inner ring seems to be of glass, set in a holder of soft nickel or brass, and it cracks very easily.

The fly rod's top guide, the "tip top," is of even more importance than the butt guide in resisting wear, for the line, of course, extends from it at more or less a right angle when the rod is in use.

The "pear-shape" tip top is now almost universally used, and is excellent in either the tungsten or chrome-plated style; and the latter again will wear longer. The size of the ring will be governed by the tube size, indicated by its inside diameter, such as $\frac{5}{64}$ inch, suitable for many fly rods, and the diameter of the wire will correspond to insure a well-balanced fitting. These tip tops are built without struts or braces from the barrel to the ring, as are the larger ones for heavy rods. It is important to add just as little weight as possible to the fly rod tip. Again, agate tip tops, while handsome and very smooth, are easily subject to breakage resulting in a ruined or injured fly line.

SPACING FLY ROD GUIDES.

The length of the rod—and your personal preference somewhat—will determine the position of the butt guide. Usually it is placed from 32 to 34 inches from the end of the butt on a 9-foot rod. This distance is practical, for it permits a long angle of the line and so less friction as it is stripped out by the hand when casting. In a three-piece rod, also, this spacing of the guide places it near the lower ferrule, and so equalizes some of the strain at that point, the line being bridged over it. In shorter rods the butt guide will then be placed perhaps 28 or 30 inches from the butt.

The total number of guides and their size will also be governed by the rod length and calibre, at least to some extent. Your own preference may be considered here. Some men, for instance, like large guides, and less of them. Others want plenty of guides, even at the expense of loading down a slender stick. It is true that a maximum number of snake guides—and their total weight is not great—will add some strength to a rod under great stress, by more evenly distributing the strain through the entire arc. To better study this point we need only to run the line through a butt guide, and then through two or three widely spaced snake guides up to the tip top. More strain is imposed on the arching stick, and a great deal more friction occurs at each guide.

A happy medium seems best. The usual practice is to use twelve guides on a 9-foot rod, exclusive of the tip top, and many of the finest custom-built rods will be so fitted. The same number of guides will also work well on a rod of 9½ feet, though one more might be employed. On a three-piece rod the butt section will have the stripping guide alone, the middle section five snake guides and the tip six. An 8½-foot rod does nicely with four guides on the middle section and five on the tip. On shorter rods the number of guides can be correspondingly reduced.

Good average spacing—for trial—on the nine-footer of "medium" action would be, starting from the butt: stripping guide, 33 inches; middle section: 8, 8, 7½, 7¼, 7 inches, the fifth guide being almost to the female ferrule of the middle joint, again bridging the ferrule. Tip: from this last guide, 6½, 6¼, 6, 5¾, 5 and 4½ inches. Distance of the last guide from the tip top then is about 4 inches. This arrangement will probably come close to fitting the arc of the rod under stress, and so equalize the strain. The varying taper of the rod, however, may necessitate a slightly different spacing for best advantage.

Similar spacing of guides will apply to the two-piece rod, having only one ferrule, although two snake guides will usually fit on the butt section above the stripping guide, the last coming quite close to the female ferrule and having the advantage noted above.

The final and best test for guide spacing is to place them about as indicated and bind one foot of each guide with a strip of Scotch tape to hold it in place, then assemble the rod and run the line through the guides, with the end fastened some distance from you. Considerable flexing now brings the line down from its slight elevation above the rod—it being held guides uppermost—until it touches the wood or even goes below the top surface. A bit of study now will indicate how a bit of shifting may equalize, and distribute the strain, which is indicated by a sharp arc toward the tip.

Again, for this average 9-foot rod mentioned above, we may select the guide sizes something as follows: middle section, two guides size No. 2 and three guides size No. 1; tip section one No. 1 guide, three No. 1-0 size and two No. 2-0. Again, on a quite powerful middle section all five guides can be as large as No. 2, which will help render a free-running line. Tip guides may be a size larger, also, if kept within a reasonable size and weight in proportion to the tip.

The smaller rod, perhaps a very light 7½ footer, will require a guide spacing somewhat similar, but smaller guides throughout will be in order. While the few grains' weight of, say, five little snake guides seems infinitesimal, the increase to a considerably larger size can at once be noted in the action of a very slender tip. The spacing, too, is particularly important on these little rods, and will have considerable effect on the rod's performance.

An even more practical test, if you care to take the trouble, is to bind all guides securely with the Scotch tape—it will hold—line up the rod and go out and actually cast with it. This is best done with a leader attached so that the test really simulates fishing conditions. You may find the line will strip better with the butt guide shifted a bit; and that the rod will sometimes perform better after the tip guides are shifted slightly. For actual casting, the tip top should be cemented into place.

A fly rod hook ring, for holding the hook or fly when the rod is carried fully rigged, is a handy little affair. It is usually a small ring, with a flat keeper which is wrapped on just forward of the hand grasp. The smallest size snake guide, mounted in similar fashion, also does very well.

FITTINGS FOR CASTING RODS.

On all rods, such as bass and spinning rods, which are used to cast "from the reel," the guides should be of good size although within a reasonable limit, governed by the size of the stick. In this style of casting, friction is also important, both as to line wear and one's ability to cast smoothly. Ring guides are always best, and again the steel, chrome-plated type is indicated. Fine agate ring guides are to be found on some high-grade rods for bait casting and they work beautifully, but are always subject to cracking or breaking.

The light, 6-foot bass rod will do nicely with five guides, which I believe should be a minimum for this length. The butt guide may go about 30 inches up from the rod butt, and the remaining ones spaced as follows: 10, 9½, 9 and 8 inches, the last guide being about 6½ inches from the tip top. Again a test with line through the guides will indicate any change in spacing which may be required by the taper of the stick. The $^{14}/_{64}$ size, inside ring diameter, will be about right for the two lower guides, and the $^{12}/_{64}$ or even $^{10}/_{64}$ for the three upper guides. Again, use guides as large as possible without overloading the upper part of the rod.

The spinning rod may be classed as a casting rod, although it is far more pliant for a given length than the conventional bait casting rod. This necessitates more guides as does its length, usually from 7 to 8 feet. The butt guide on this rod should be very large, ⅞ inch ring diameter being a practical size, to permit easy flow of the line as it spirals out from the fixed spool of the spinning reel. The remaining guides, usually five in number for a 7½-foot rod, are then spaced about: 11½, 10, 9, 7½ and 6 inches apart, the last being 5½ inches from the end of the tip top.

Special, very light wire, chrome-plated guides are available for spinning rods, permitting a large diameter throughout without excessive weight to disturb the action of this little tool, which flexes almost like a fly rod during the cast. These, too, may well be mounted on the light-action bait casting rod, such as the very whippy tournament rod used for ⅜-ounce casting plugs.

A new and interesting design of light ring guide is being manufactured by the Berner Company of Los Angeles, California. This nicely finished and shaped guide is called the "Foulproof," and thus features a long, tapering "rear leg" along which the line slides without any tendency to loop as it enters the ring. The material is spring-tempered monel metal, chrome-plated, and should endure a lifetime of wear. This guide should prove ideal for the spinning and bait casting rod and has possibilities for salt water rods, being made in all sizes up to ⅞ inch inside ring diameter.

Tip tops for bait casting and spinning rods should always be chrome-plated for long wear and have a ring of ample size, for here again, friction is detrimental to good casting and causes line wear. One type of tip top has a large ring soldered into a strong "fork" which, in turn, is soldered into the barrel. But there is a tendency sometimes for the line to foul and loop around this fitting. The most practical type has two strong struts bracing the ring to the barrel, being hard-soldered at each point.

These tip tops come in barrel sizes—inside diameter—from 5/64 inch up to those suitable for large salt water rods, and are almost universally used on all types of rods which employ a bait casting or spinning reel, where considerable friction is set up by the speed of the line through the guides. It is well to remember, too, that while line friction is important in regard to the fly rod, the fly line is smoothly dressed and flows freely. In bait casting and salt water fishing an undressed line of linen, silk or nylon is used. These, while smoothly braided in modern practice, still do set up strong friction as the raw line speeds through the guides.

FITTINGS FOR SALT WATER RODS.

Here large ring guides are essential, and we almost might say "the larger the better." Most rods of this type, of husky calibre, will not be affected by a bit of extra weight. The lighter rod tips in some instances, however, such as the California Tuna Club's "3–6" stick, may require that total weight of guides be considered to keep the entire tip, fully fitted, under the maximum weight of four ounces. If you are building such a rod, or the six-ounce "Light-Tackle stick," it is well to keep this weight regulation in mind and design the rod to have a margin of weight to permit practical and ample guides.

Heavy steel guides, chrome-plated, are in general use for salt water rods and are excellent in either the Perfection or Mildrum pattern. The latter style is a bit heavier for a given size than the Perfection and, in my experience, has seemed to have a thinner coating of the glass-hard chrome.

Very good stainless steel guides also are available for these large rods. They are strongly built, well designed, with long feet for secure wrapping, and the rings are extremely hard. This material will be entirely unaffected by salt water and may prove even better than the chrome finish in resisting corrosion.

Salt water sticks used for boat fishing, such as "light," "medium" and "heavy-tackle" rods, usually being about 5 feet long, should have at least five guides, and six are often better. The butt guide is usually placed about 24 inches from the end of the ferrule and when five guides are used, they are spaced about as follows: 10, 9, 7, 7 inches apart, the last being from 6 to 7 inches from the tip top.

Special bridge-type ring guides are often mounted on these heavy-duty rods. This fitting has a frame which carries the ring considerably above the stick and so elevates the line well above the rod, thus preventing rubbing along the wood when the rod is sharply arched. Thus, some advantage may be gained, but possibly at the expense of the rod twisting or turning due to the higher point of strain in relation to the rod's axis.

Tip tops for all sea rods should have large rings, strong struts and be always chrome-plated or of stainless steel. A new and very fine fitting is now made of stainless steel, with the *ring* of beautifully polished carboloy, a marvelous material almost diamond-hard and as smooth as the finest agate. All these large tip tops require careful fitting to the wood before cementing, since the bearing surface is comparatively short, is somewhat tapered, and great strain is imposed on the tip when playing a large fish.

I have purposely refrained from advising guide size and spacing for surf rods, since these sticks vary greatly in length and calibre and are used with a wide variety of casting weights. The "dyed in the wool" lover of this great sport will have his own ideas as to guides and accept none other. The rule, however, is to place a large butt guide well up the rod and employ only one or two more between it and the tip top.

Large agate guides are often still favored for this rod, since it will not undergo the same kind of rough treatment as boat rods, and so the danger of breaking the agate rings is reduced. Certainly the very smoothest possible guides are best in this game, where extreme distance in casting, with a minimum of friction is all-important.

CHAPTER 17

WRAPPING AND FINAL FINISHING

While a nice varnish finish for the bamboo rod has long been the accepted practice and does provide a protective coat against abrasion, there is some question as to whether ordinary rod varnish is thoroughly waterproof. Thus, some experimenting has been done with an oil finish on the wood, the silk windings, only, being varnished to protect them against wear. At least one maker of fine rods in the past used a method of oil finishing, and I believe that one or more manufacturers still employ it.

Some competent authorities, among them Robert Crompton, have contended that a heavy varnish coat will considerably affect and slow up the fine action of a fly rod, since a more or less unyielding sleeve of varnish encases the wood. My own modest experiments, particularly on very light action rods, seem to bear out these findings. Thus, for light rods, where the natural action of the bamboo should not be impeded in any manner other than by the necessary ferrules, an oil finish may interest you.

The kinds of oils used commercially and the methods of application are probably trade "secrets." The treatment may involve special apparatus, require considerable time and possibly the use of at least a partial vacuum to force the oil well into the wood. Such methods will be beyond the worker in his home shop, but a simple application is possible. The light "floor sealer," used on hardwood floors before final finishing, is a very thin liquid which penetrates well. Another preparation, "Lion Oil," sold for wood treatment, and to be found in hardware and paint stores, is also good when thinned with turpentine in about equal proportions.

Either of these oils can be applied to the finished and ferruled sticks with a rag or small brush and set away to dry. Another coat, or perhaps two, will seem to penetrate the wood partially, and then remain as a sticky film on the surface. After a day or two this can be buffed off with fine steel wool and a coarse rag, leaving the wood with a soft, velvety finish which may be brought up to a bit more lustre by a light application of wax. I have purposely subjected a fly rod so finished to frost, dew and rain, without later wiping or drying it, and so far it shows no ill effect from the moisture.

It would seem wise, however, to use the oil finish only when the rod is glued with thoroughly waterproof glue, such as one of the synthetic resins, which will withstand actual soaking without harm; and I am not sure this would be a safe procedure if you have used the otherwise reliable caseins, although they are highly water-resistant.

There is much to be said in favor of a fine varnish finish and, after all, it does produce a handsome rod, as well as protect it a great deal from moisture—and from abrasion in use. Scratches and small dents can always be touched up with varnish again, after a bit of smoothing with a fine abrasive. Severe marring can penetrate the outer fibres of oil finished bamboo, which really has no protective coating, so the choice of finish is yours. If varnish is used in a moderate amount on the light fly rod the action certainly will not be greatly affected.

Now we are ready for the wrapping, regardless of the final finish employed. With the guides properly spaced, as described in the preceding chapter, and each held in place by a strip of Scotch tape around one foot, sight down the rod sticks and line them up by slightly shifting any that require it.

WRAPPING FLY RODS.

Good rod-wrapping silk, sold by supply houses and sporting goods stores, is better than ordinary sewing silk, being stronger and having much of the fuzz removed. Size No. 00 is best for fly rods and makes a neater job than size A which is often used. Nylon wrapping thread is very strong, but does not seem to be made in sizes smaller than "A." Also, it stretches somewhat when being applied and is a bit more difficult to handle. Either material comes in almost any desired color.

Here your ideas as to color and general ornamentation of the rod may run to any limit within reason, from a barber pole effect, achieved by contrasting main winds and trim, to the modest and usually pleasing method of using a tan, light brown or yellow thread much the same color as the bamboo. A subdued but fancier finish is possible with a light color for the main winds, each edged with a narrow one of contrasting shade. Tan with a narrow red trim is quite effective; also gold-yellow trimmed with red or black.

In any case, these narrow trim winds applied snugly at the outer ends of the main winds on guides and ferrules should be made with not more than six or seven turns of thread, or the finished job will appear decidedly amateurish.

Choice of colors should be governed somewhat by how the thread will be finished. If varnish is applied directly to the wrappings later, it will darken the thread several shades, affecting silk more than it does nylon. Thus, light red becomes maroon or cardinal, and often appears dingy. Tan turns quite brown and a dark brown an almost black, unpleasant color. The light yellows and golden shades seem least affected by the varnish. Most of the jasper threads, composed of two pleasing colors, such as black and grey and black and tan, also will be changed for the worse.

The varnishing of the wrappings, along with the entire rod, is a more practical method than using color preservative on them first. The varnish penetrates the thread and glues it to the wood, the final coats building up a protective finish. This straight-varnish finishing will endure for many years, while the use of preservative, which is usually clear lacquer, sometimes binds the threads together rather than to the wood, forming a kind of sleeve which may loosen and come off when the thread is abraded in rough use.

Again, as to method, the choice is yours. I personally like to varnish directly over wrappings on heavy-duty rods which are bound to endure hard service, and to preserve the natural color of silk on lighter fly rods. After all, the "fairy" fly rod is a delicate instrument, to be used with appropriate reverence, and certainly we may be permitted the full range of esthetic choice as to color of the windings and the finish.

Some anglers, and most professional craftsmen, I believe, employ some mechanical means of holding and spinning the rod stick while the thread is applied. But I do not think this is necessary. In every phase of home rod work, simplicity plus a bit of ingenuity will be best.

The rod stick, up to a length of 4 or 5 feet, can be held under the arm against the body, with the thumb and index finger turning it, while the other hand feeds on the thread and maintains the proper tension. It would be impossible to apply the thread snugly if more speed were used than is possible by this method, and I have not found a mechanical tension control of the thread at all necessary. The spool revolves on a pin set in a hardwood block, and the feeding hand soon "learns" just how much tension to apply.

If this hand is kept at least 6 inches from the stick, it is possible to keep the thread at an angle a bit more than at right angles to the rod, thus crowding the thread until it seems to almost climb over the wraps, and so lay close. The tiniest spaces must be watched for and the thread either unwrapped for correction or the thumb nail used to press it snugly together. This can be accomplished without changing the position of the fingers which hold the stick.

All the guides, if not nicely tapered at the end of each foot, should have been ground or filed down to a thin edge at this point, to permit the thread starting from the wood and over the end without a noticeable step. The thread is started about two turns out from the end of the foot, with an inch or so of its length turned back along the rod flat. The thumb which revolves the rod is placed over these first turns until friction increases the tension, and so renders holding unnecessary. After six or eight turns, the underlying thread is cut off against the wood with a razor blade.

The wrapping should finish just before the foot of the guide starts to rise above the rod. Here the pull-under loop is the best method of finishing the job. A few lengths of the thread about 3 inches long should be laid out in advance for these loops. About six turns from the stopping place the looped piece is laid along the rod, both ends projecting, and wrapped over. The thread is then cut, leaving about a 3-inch end which is slipped through the loop, which in turn is pulled back under the tight wrappings, and entirely *out,* bringing the end of the wrapping thread with it. This is cut off flush with a sharp razor blade, and we have a truly invisible and tight finish.

The illustrations show the various above steps and also how to finish a narrow wrap without use of the loop. This, as employed for narrow "trim" wraps and for intermediate ones along the rod's length, usually has only six or seven complete turns of thread, so all are placed over the first end. This is laid back, forming a loop in itself, after three turns, and wound over. The cut end of the thread from the spool then is run through the loop and drawn back to the center of the wind, to be trimmed flush before the loop appears, while the other end, which has disappeared under the finish of the wrap is cut close up.

If contrasting trim is used, it is better left until all utility winding is finished, to save changing thread and method so many times; intermediate windings likewise. These are not necessary and are seldom found on fine fly rods these days, but they may be used in a narrow style and spaced from 1 to 2 inches apart, the exact division depending somewhat on how far apart the guides are. Their advantages—which are doubtful at best on a good, light rod—will be more than offset by the extra work in

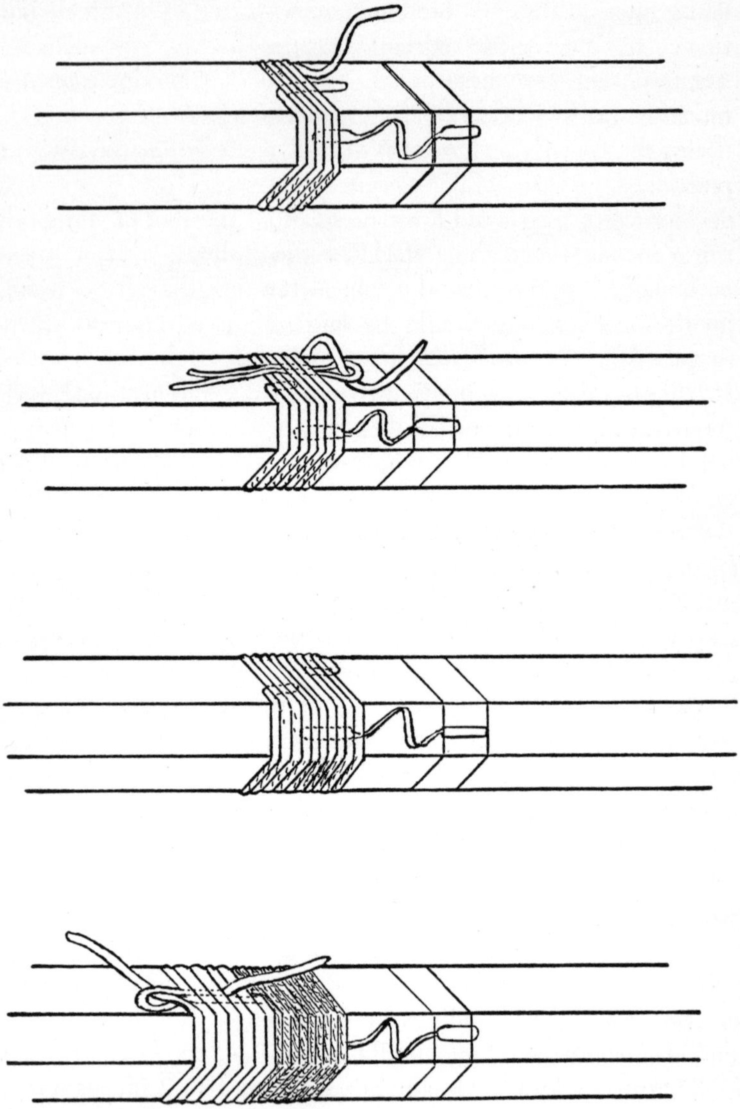

Narrow Trim Wind

Method used for wrapping guides and ferrule ends.

doing them neatly, and by their tendency to slow up the action somewhat when many are used, particularly on a slender tip. Thus, if they are placed *very* close together, say at half-inch intervals, the rod may become somewhat logy.

But this is your rod, to be finished in any style you like, so probably long ago, while waiting for the sticks to season, you decided all the details of this important affair. The principal thing now is to do the job well. Amateurish work, so often noted on home-finished rods, need not mar your finished rod. So watch for the little tell-tale gaps in the thread and employ that handy thumb nail.

After one foot of all the guides on a section has been wrapped, again sight down the stick, and realign any which may be slightly out by pushing strongly against the free foot. Then the final guide windings may be completed.

Practically all good ferrules have tapered ends over which the thread starts smoothly, and a narrow shoulder turned on the metal for a stopping point. Start the thread only a turn or two from the metal and tie off with the loop snug against this shoulder. It is usually best to follow this first wrapping with another over it, this being particularly necessary when using serrated or split ferrules which may permit the ferrule cement to come through a single winding after varnish is applied and so stain the silk. If the thread is started many turns down on the wood below the ferrule end, it will nearly always break or separate later when the rod is in use, due to the flexing of the wood at the end of the ferrule.

Sometimes, when the serrated ends of the ferrule are slightly sprung upward and do not lie close to the wood, the winding process may have to be reversed: the thread started at the shoulder of the ferrule and wrapped down over the end, thus binding the serrated sections tightly as it progresses. It is difficult, however, to make the final wrap from metal to wood without an unsightly gap.

The narrow, ornamental trim windings are next in order, and

Buy two ox-hair sign painter's brushes, each a quarter inch wide, and keep one for the lacquer work, the other for varnish. This is the best type of brush I have found, being fairly stiff, yet with fine hair, and far more suitable than camel's hair or sable brushes, which are too soft.

Use the brush edgewise in going around the windings alone, to keep the lacquer off the wood. Three coats will be required to fully protect the color, and these should each have about an hour to dry before applying the next. The lacquer will usually need thinning about "half and half" for best results. The thinner, of course, will clean your brush and this should be done each time immediately after it is used.

If the windings are to have varnish alone, use the other brush, and in the same manner. At least twenty-four hours will be needed for drying between coats, and perhaps two full days. In going around the windings it is handy to rest one end of the stick in a shallow hole drilled slanting in a block of wood, and revolve the stick, held at the other end, against the brush.

If the rod is to be fully varnished the wrappings will need only the three color preserving coats of lacquer first. If this is not used it is best to give the silk one extra coat of varnish first before going over the entire rod. This will sink well in, and permit the next coat to build up over the surface.

Occasionally good rod varnish is sold in the tackle stores, and I have found both South Bend and Heddon's of fair quality. The household varnish called Sapolin, and sold in small cans, usually seems excellent. Also, the product of the Seal-tite Company of Los Angeles is very fine. It comes in small bottles in a quantity enough for about three rods. It is a very light-bodied and *light-colored* varnish, both these characteristics being essential for rod work.

Here, if you want a glassy-smooth, professional varnish job, resolve to be "clean," and an absolute crank about dust. Do not permit company nor answer the telephone after you start the job, for all movement raises dust and lint in the air. It is these tiny particles, invisible to the eye unless seen against a strong light, which make the lumps we see in most amateur work.

Work in a small room without draft and even go so far as to remove your shirt to prevent lint coming from it. First wipe the rod sticks with a bit of cloth dipped in benzine, to remove the grease from finger marks. Then dust each carefully, full length and around the wrappings with a small, *clean* paint brush, from which you have removed all dust by repeated ruffling of the bristles. Stand the sticks along the back of the work table or bench while you wipe its surface carefully with a damp cloth.

The varnish is best poured into a small cup or bowl which also has been carefully dusted; and a small quantity at a time thus used will prevent its thickening when long exposed to the air. Then, work the dry varnish brush with the finger tip to be sure it is entirely free of dust, and you are ready. Short rod sections should be held at an angle by the ferrule alone; longer ones you may grasp at a wrapping. If the varnish seems thick, thin with a bit of turpentine.

Fill the brush full of varnish, wipe off the surplus beyond the dripping point, and first go around the silk on the lower ferrule and the next guide, then through under the bridge of the guide. Then brush along the flats from one winding to the next, taking each in turn. Usually with a small rod, your quarter-inch brush will lap over to do two flats at once quite neatly. Regardless of some contention that the first coat should be very thin and brushed in, I find that this usually makes a patchy-looking second coat. It is better to thin the varnish to flow on easily and evenly. The method of flowing with an always full brush is best.

It is quite easy, with a bit of practice, to go from one guide to the next without lifting the brush or going back over the strokes. This insures a good, finished job later. If the room is kept at a temperature of 75 degrees or higher the varnish will work better. The varnished sticks may be left standing along the back of your table if the room can be kept closed and unused for a day; otherwise try to find a closet or cupboard in which they may be stood or hung.

Most varnish will dry in twenty-four hours in good weather, but will not be actually hard enough for a second coat in less than two days, and perhaps three. If the surface feels at all tacky it is better to wait still another day. No. 00 steel wool is probably best for rubbing down the gloss for the next coat and is far easier to use than pumice stone and water which is sometimes advised and which is very difficult to remove from around and under the guides.

Use a small pad of the wool held over the end of the finger, and work from guide to guide, along the flats, being careful not to bear hard on the corners and so remove any varnish. The surface need be only dulled to give the next coat a good hold.

Three full coats of varnish will be ample for a fly rod, and more will simply build up a sleeve which can impede the rod's fine action. This thick sleeve will also chip more easily than if less varnish is used. Rub down the second coat as you did the first, and apply the dust brush *diligently* to remove dust, lint and the tiny bits of steel wool. The brush should be washed thoroughly in turpentine immediately after using, then scrubbed with soap and water and put to dry in a drawer or cabinet. Thus cared for, it will last for years.

Some anglers like to rub down the final varnish coat with pumice, then wax the rod for a soft, velvet finish. This procedure is entirely optional and at least will do no harm.

WRAPPING BAIT CASTING AND HEAVIER RODS.

The spinning rod, while essentially a bait casting tool, has almost the flexibility of the fly rod, and it is light in weight. So No. 00 wrapping silk of your choice, or the slightly heavier No. A thread will be suitable. Intermediate windings, as on the fly rod, will be optional. The method of wrapping and finishing will be exactly the same as for the fly rod.

Other bait casting rods, sometimes undergoing hard service

when used in boats and scarred by many-hooked plugs, will do better with No. A wrapping thread which will stand much more abrasion than the lighter size without separating. Size A nylon is very strong, it lies very close if applied tightly, and is quite suitable for these rods.

The lighter rods of this type will have ample protection with three varnish coats; but those for boat use, with the inevitable scuffing and marring of the finish, will be better protected with four, or even five, fairly heavy coats. I do not think that even the most critical admirer of unhampered rod action could detect any slowing effect of ample varnish on rods heavier than the lighter type of spinning rod.

SALT WATER RODS.

These larger sticks, being usually one piece, may present a bit of trouble in wrapping. But those no longer than 5 feet, I find, can be held under the arm as are the shorter rod sections. The longer trolling and surf rods may require a bit of ingenuity on your part, in rigging up some form of temporary jig to support and render easy the turning of the stick. I have seen good work done with the rod supported on two pairs of spools, mounted on axles of horizontal nails, while a heavy book placed on the wrapping thread supplied the necessary tension.

The larger rods should be wrapped with thread of at least size "E." This, in nylon, is extremely strong and, when well varnished, will resist severe wear for a very long period. Here, again, the plan of varnishing directly over the thread, without use of lacquer, has great merit; this treatment literally "glues" the thread and almost *welds* it into the rod.